T0275437

Lakes: A Very Short Introduction

VERY SHORT INTRODUCTIONS are for anyone wanting a stimulating and accessible way into a new subject. They are written by experts, and have been translated into more than 45 different languages.

The series began in 1995, and now covers a wide variety of topics in every discipline. The VSI library currently contains over 500 volumes—a Very Short Introduction to everything from Psychology and Philosophy of Science to American History and Relativity—and continues to grow in every subject area.

Very Short Introductions available now:

ACCOUNTING Christopher Nobes
ADOLESCENCE Peter K. Smith
ADVERTISING Winston Fletcher
AFRICAN AMERICAN RELIGION Eddie S. Glaude Jr
AFRICAN HISTORY John Parker and Richard Rathbone
AFRICAN RELIGIONS Jacob K. Olupona
AGEING Nancy A. Pachana
AGNOSTICISM Robin Le Poidevin
AGRICULTURE Paul Brassley and Richard Soffe
ALEXANDER THE GREAT Hugh Bowden
ALGEBRA Peter M. Higgins
AMERICAN HISTORY Paul S. Boyer
AMERICAN IMMIGRATION David A. Gerber
AMERICAN LEGAL HISTORY G. Edward White
AMERICAN POLITICAL HISTORY Donald Critchlow
AMERICAN POLITICAL PARTIES AND ELECTIONS L. Sandy Maisel
AMERICAN POLITICS Richard M. Valelly
THE AMERICAN PRESIDENCY Charles O. Jones
THE AMERICAN REVOLUTION Robert J. Allison
AMERICAN SLAVERY Heather Andrea Williams
THE AMERICAN WEST Stephen Aron
AMERICAN WOMEN'S HISTORY Susan Ware
ANAESTHESIA Aidan O'Donnell
ANALYTIC PHILOSOPHY Michael Beaney
ANARCHISM Colin Ward
ANCIENT ASSYRIA Karen Radner
ANCIENT EGYPT Ian Shaw
ANCIENT EGYPTIAN ART AND ARCHITECTURE Christina Riggs
ANCIENT GREECE Paul Cartledge
THE ANCIENT NEAR EAST Amanda H. Podany
ANCIENT PHILOSOPHY Julia Annas
ANCIENT WARFARE Harry Sidebottom
ANGELS David Albert Jones
ANGLICANISM Mark Chapman
THE ANGLO-SAXON AGE John Blair
ANIMAL BEHAVIOUR Tristram D. Wyatt
THE ANIMAL KINGDOM Peter Holland
ANIMAL RIGHTS David DeGrazia
THE ANTARCTIC Klaus Dodds
ANTISEMITISM Steven Beller
ANXIETY Daniel Freeman and Jason Freeman
THE APOCRYPHAL GOSPELS Paul Foster
ARCHAEOLOGY Paul Bahn
ARCHITECTURE Andrew Ballantyne
ARISTOCRACY William Doyle
ARISTOTLE Jonathan Barnes
ART HISTORY Dana Arnold
ART THEORY Cynthia Freeland

ASIAN AMERICAN HISTORY Madeline Y. Hsu
ASTROBIOLOGY David C. Catling
ASTROPHYSICS James Binney
ATHEISM Julian Baggini
THE ATMOSPHERE Paul I. Palmer
AUGUSTINE Henry Chadwick
AUSTRALIA Kenneth Morgan
AUTISM Uta Frith
THE AVANT GARDE David Cottington
THE AZTECS Davíd Carrasco
BABYLONIA Trevor Bryce
BACTERIA Sebastian G. B. Amyes
BANKING John Goddard and John O. S. Wilson
BARTHES Jonathan Culler
THE BEATS David Sterritt
BEAUTY Roger Scruton
BEHAVIOURAL ECONOMICS Michelle Baddeley
BESTSELLERS John Sutherland
THE BIBLE John Riches
BIBLICAL ARCHAEOLOGY Eric H. Cline
BIG DATA Dawn E. Holmes
BIOGRAPHY Hermione Lee
BLACK HOLES Katherine Blundell
BLOOD Chris Cooper
THE BLUES Elijah Wald
THE BODY Chris Shilling
THE BOOK OF MORMON Terryl Givens
BORDERS Alexander C. Diener and Joshua Hagen
THE BRAIN Michael O'Shea
BRANDING Robert Jones
THE BRICS Andrew F. Cooper
THE BRITISH CONSTITUTION Martin Loughlin
THE BRITISH EMPIRE Ashley Jackson
BRITISH POLITICS Anthony Wright
BUDDHA Michael Carrithers
BUDDHISM Damien Keown
BUDDHIST ETHICS Damien Keown
BYZANTIUM Peter Sarris
CALVINISM Jon Balserak
CANCER Nicholas James
CAPITALISM James Fulcher
CATHOLICISM Gerald O'Collins
CAUSATION Stephen Mumford and Rani Lill Anjum
THE CELL Terence Allen and Graham Cowling
THE CELTS Barry Cunliffe
CHAOS Leonard Smith
CHEMISTRY Peter Atkins
CHILD PSYCHOLOGY Usha Goswami
CHILDREN'S LITERATURE Kimberley Reynolds
CHINESE LITERATURE Sabina Knight
CHOICE THEORY Michael Allingham
CHRISTIAN ART Beth Williamson
CHRISTIAN ETHICS D. Stephen Long
CHRISTIANITY Linda Woodhead
CIRCADIAN RHYTHMS Russell Foster and Leon Kreitzman
CITIZENSHIP Richard Bellamy
CIVIL ENGINEERING David Muir Wood
CLASSICAL LITERATURE William Allan
CLASSICAL MYTHOLOGY Helen Morales
CLASSICS Mary Beard and John Henderson
CLAUSEWITZ Michael Howard
CLIMATE Mark Maslin
CLIMATE CHANGE Mark Maslin
CLINICAL PSYCHOLOGY Susan Llewelyn and Katie Aafjes-van Doorn
COGNITIVE NEUROSCIENCE Richard Passingham
THE COLD WAR Robert McMahon
COLONIAL AMERICA Alan Taylor
COLONIAL LATIN AMERICAN LITERATURE Rolena Adorno
COMBINATORICS Robin Wilson
COMEDY Matthew Bevis
COMMUNISM Leslie Holmes
COMPLEXITY John H. Holland
THE COMPUTER Darrel Ince
COMPUTER SCIENCE Subrata Dasgupta
CONFUCIANISM Daniel K. Gardner
THE CONQUISTADORS Matthew Restall and Felipe Fernández-Armesto
CONSCIENCE Paul Strohm
CONSCIOUSNESS Susan Blackmore
CONTEMPORARY ART Julian Stallabrass

CONTEMPORARY FICTION Robert Eaglestone
CONTINENTAL PHILOSOPHY Simon Critchley
COPERNICUS Owen Gingerich
CORAL REEFS Charles Sheppard
CORPORATE SOCIAL RESPONSIBILITY Jeremy Moon
CORRUPTION Leslie Holmes
COSMOLOGY Peter Coles
CRIME FICTION Richard Bradford
CRIMINAL JUSTICE Julian V. Roberts
CRITICAL THEORY Stephen Eric Bronner
THE CRUSADES Christopher Tyerman
CRYPTOGRAPHY Fred Piper and Sean Murphy
CRYSTALLOGRAPHY A. M. Glazer
THE CULTURAL REVOLUTION Richard Curt Kraus
DADA AND SURREALISM David Hopkins
DANTE Peter Hainsworth and David Robey
DARWIN Jonathan Howard
THE DEAD SEA SCROLLS Timothy H. Lim
DECOLONIZATION Dane Kennedy
DEMOCRACY Bernard Crick
DEPRESSION Jan Scott and Mary Jane Tacchi
DERRIDA Simon Glendinning
DESCARTES Tom Sorell
DESERTS Nick Middleton
DESIGN John Heskett
DEVELOPMENTAL BIOLOGY Lewis Wolpert
THE DEVIL Darren Oldridge
DIASPORA Kevin Kenny
DICTIONARIES Lynda Mugglestone
DINOSAURS David Norman
DIPLOMACY Joseph M. Siracusa
DOCUMENTARY FILM Patricia Aufderheide
DREAMING J. Allan Hobson
DRUGS Les Iversen
DRUIDS Barry Cunliffe
EARLY MUSIC Thomas Forrest Kelly
THE EARTH Martin Redfern
EARTH SYSTEM SCIENCE Tim Lenton
ECONOMICS Partha Dasgupta
EDUCATION Gary Thomas
EGYPTIAN MYTH Geraldine Pinch
EIGHTEENTH-CENTURY BRITAIN Paul Langford
THE ELEMENTS Philip Ball
EMOTION Dylan Evans
EMPIRE Stephen Howe
ENGELS Terrell Carver
ENGINEERING David Blockley
ENGLISH LITERATURE Jonathan Bate
THE ENLIGHTENMENT John Robertson
ENTREPRENEURSHIP Paul Westhead and Mike Wright
ENVIRONMENTAL ECONOMICS Stephen Smith
ENVIRONMENTAL LAW Elizabeth Fisher
ENVIRONMENTAL POLITICS Andrew Dobson
EPICUREANISM Catherine Wilson
EPIDEMIOLOGY Rodolfo Saracci
ETHICS Simon Blackburn
ETHNOMUSICOLOGY Timothy Rice
THE ETRUSCANS Christopher Smith
EUGENICS Philippa Levine
THE EUROPEAN UNION John Pinder and Simon Usherwood
EUROPEAN UNION LAW Anthony Arnull
EVOLUTION Brian and Deborah Charlesworth
EXISTENTIALISM Thomas Flynn
EXPLORATION Stewart A. Weaver
THE EYE Michael Land
FAIRY TALE Marina Warner
FAMILY LAW Jonathan Herring
FASCISM Kevin Passmore
FASHION Rebecca Arnold
FEMINISM Margaret Walters
FILM Michael Wood
FILM MUSIC Kathryn Kalinak
THE FIRST WORLD WAR Michael Howard
FOLK MUSIC Mark Slobin
FOOD John Krebs
FORENSIC PSYCHOLOGY David Canter
FORENSIC SCIENCE Jim Fraser
FORESTS Jaboury Ghazoul
FOSSILS Keith Thomson
FOUCAULT Gary Gutting

THE FOUNDING FATHERS R. B. Bernstein
FRACTALS Kenneth Falconer
FREE SPEECH Nigel Warburton
FREE WILL Thomas Pink
FREEMASONRY Andreas Önnerfors
FRENCH LITERATURE John D. Lyons
THE FRENCH REVOLUTION William Doyle
FREUD Anthony Storr
FUNDAMENTALISM Malise Ruthven
FUNGI Nicholas P. Money
THE FUTURE Jennifer M. Gidley
GALAXIES John Gribbin
GALILEO Stillman Drake
GAME THEORY Ken Binmore
GANDHI Bhikhu Parekh
GENES Jonathan Slack
GENIUS Andrew Robinson
GEOGRAPHY John Matthews and David Herbert
GEOPOLITICS Klaus Dodds
GERMAN LITERATURE Nicholas Boyle
GERMAN PHILOSOPHY Andrew Bowie
GLOBAL CATASTROPHES Bill McGuire
GLOBAL ECONOMIC HISTORY Robert C. Allen
GLOBALIZATION Manfred Steger
GOD John Bowker
GOETHE Ritchie Robertson
THE GOTHIC Nick Groom
GOVERNANCE Mark Bevir
GRAVITY Timothy Clifton
THE GREAT DEPRESSION AND THE NEW DEAL Eric Rauchway
HABERMAS James Gordon Finlayson
THE HABSBURG EMPIRE Martyn Rady
HAPPINESS Daniel M. Haybron
THE HARLEM RENAISSANCE Cheryl A. Wall
THE HEBREW BIBLE AS LITERATURE Tod Linafelt
HEGEL Peter Singer
HEIDEGGER Michael Inwood
THE HELLENISTIC AGE Peter Thonemann
HEREDITY John Waller
HERMENEUTICS Jens Zimmermann
HERODOTUS Jennifer T. Roberts
HIEROGLYPHS Penelope Wilson
HINDUISM Kim Knott
HISTORY John H. Arnold
THE HISTORY OF ASTRONOMY Michael Hoskin
THE HISTORY OF CHEMISTRY William H. Brock
THE HISTORY OF CINEMA Geoffrey Nowell-Smith
THE HISTORY OF LIFE Michael Benton
THE HISTORY OF MATHEMATICS Jacqueline Stedall
THE HISTORY OF MEDICINE William Bynum
THE HISTORY OF TIME Leofranc Holford-Strevens
HIV AND AIDS Alan Whiteside
HOBBES Richard Tuck
HOLLYWOOD Peter Decherney
HOME Michael Allen Fox
HORMONES Martin Luck
HUMAN ANATOMY Leslie Klenerman
HUMAN EVOLUTION Bernard Wood
HUMAN RIGHTS Andrew Clapham
HUMANISM Stephen Law
HUME A. J. Ayer
HUMOUR Noël Carroll
THE ICE AGE Jamie Woodward
IDEOLOGY Michael Freeden
THE IMMUNE SYSTEM Paul Klenerman
INDIAN CINEMA Ashish Rajadhyaksha
INDIAN PHILOSOPHY Sue Hamilton
THE INDUSTRIAL REVOLUTION Robert C. Allen
INFECTIOUS DISEASE Marta L. Wayne and Benjamin M. Bolker
INFINITY Ian Stewart
INFORMATION Luciano Floridi
INNOVATION Mark Dodgson and David Gann
INTELLIGENCE Ian J. Deary
INTELLECTUAL PROPERTY Siva Vaidhyanathan
INTERNATIONAL LAW Vaughan Lowe
INTERNATIONAL MIGRATION Khalid Koser
INTERNATIONAL RELATIONS Paul Wilkinson
INTERNATIONAL SECURITY Christopher S. Browning

IRAN Ali M. Ansari
ISLAM Malise Ruthven
ISLAMIC HISTORY Adam Silverstein
ISOTOPES Rob Ellam
ITALIAN LITERATURE
Peter Hainsworth and David Robey
JESUS Richard Bauckham
JEWISH HISTORY David N. Myers
JOURNALISM Ian Hargreaves
JUDAISM Norman Solomon
JUNG Anthony Stevens
KABBALAH Joseph Dan
KAFKA Ritchie Robertson
KANT Roger Scruton
KEYNES Robert Skidelsky
KIERKEGAARD Patrick Gardiner
KNOWLEDGE Jennifer Nagel
THE KORAN Michael Cook
LAKES Warwick F. Vincent
LANDSCAPE ARCHITECTURE
Ian H. Thompson
LANDSCAPES AND
GEOMORPHOLOGY
Andrew Goudie and Heather Viles
LANGUAGES Stephen R. Anderson
LATE ANTIQUITY Gillian Clark
LAW Raymond Wacks
THE LAWS OF THERMODYNAMICS
Peter Atkins
LEADERSHIP Keith Grint
LEARNING Mark Haselgrove
LEIBNIZ Maria Rosa Antognazza
LIBERALISM Michael Freeden
LIGHT Ian Walmsley
LINCOLN Allen C. Guelzo
LINGUISTICS Peter Matthews
LITERARY THEORY Jonathan Culler
LOCKE John Dunn
LOGIC Graham Priest
LOVE Ronald de Sousa
MACHIAVELLI Quentin Skinner
MADNESS Andrew Scull
MAGIC Owen Davies
MAGNA CARTA Nicholas Vincent
MAGNETISM Stephen Blundell
MALTHUS Donald Winch
MAMMALS T. S. Kemp
MANAGEMENT John Hendry
MAO Delia Davin
MARINE BIOLOGY Philip V. Mladenov
THE MARQUIS DE SADE John Phillips
MARTIN LUTHER Scott H. Hendrix
MARTYRDOM Jolyon Mitchell
MARX Peter Singer
MATERIALS Christopher Hall
MATHEMATICS Timothy Gowers
THE MEANING OF LIFE
Terry Eagleton
MEASUREMENT David Hand
MEDICAL ETHICS Tony Hope
MEDICAL LAW Charles Foster
MEDIEVAL BRITAIN John Gillingham
and Ralph A. Griffiths
MEDIEVAL LITERATURE
Elaine Treharne
MEDIEVAL PHILOSOPHY
John Marenbon
MEMORY Jonathan K. Foster
METAPHYSICS Stephen Mumford
THE MEXICAN REVOLUTION
Alan Knight
MICHAEL FARADAY
Frank A. J. L. James
MICROBIOLOGY Nicholas P. Money
MICROECONOMICS Avinash Dixit
MICROSCOPY Terence Allen
THE MIDDLE AGES Miri Rubin
MILITARY JUSTICE Eugene R. Fidell
MILITARY STRATEGY
Antulio J. Echevarria II
MINERALS David Vaughan
MIRACLES Yujin Nagasawa
MODERN ART David Cottington
MODERN CHINA Rana Mitter
MODERN DRAMA
Kirsten E. Shepherd-Barr
MODERN FRANCE
Vanessa R. Schwartz
MODERN INDIA Craig Jeffrey
MODERN IRELAND Senia Pašeta
MODERN ITALY Anna Cento Bull
MODERN JAPAN
Christopher Goto-Jones
MODERN LATIN AMERICAN
LITERATURE
Roberto González Echevarría
MODERN WAR Richard English
MODERNISM Christopher Butler
MOLECULAR BIOLOGY Aysha Divan
and Janice A. Royds

MOLECULES Philip Ball
MONASTICISM Stephen J. Davis
THE MONGOLS Morris Rossabi
MOONS David A. Rothery
MORMONISM Richard Lyman Bushman
MOUNTAINS Martin F. Price
MUHAMMAD Jonathan A. C. Brown
MULTICULTURALISM Ali Rattansi
MULTILINGUALISM John C. Maher
MUSIC Nicholas Cook
MYTH Robert A. Segal
THE NAPOLEONIC WARS Mike Rapport
NATIONALISM Steven Grosby
NAVIGATION Jim Bennett
NELSON MANDELA Elleke Boehmer
NEOLIBERALISM Manfred Steger and Ravi Roy
NETWORKS Guido Caldarelli and Michele Catanzaro
THE NEW TESTAMENT Luke Timothy Johnson
THE NEW TESTAMENT AS LITERATURE Kyle Keefer
NEWTON Robert Iliffe
NIETZSCHE Michael Tanner
NINETEENTH-CENTURY BRITAIN Christopher Harvie and H. C. G. Matthew
THE NORMAN CONQUEST George Garnett
NORTH AMERICAN INDIANS Theda Perdue and Michael D. Green
NORTHERN IRELAND Marc Mulholland
NOTHING Frank Close
NUCLEAR PHYSICS Frank Close
NUCLEAR POWER Maxwell Irvine
NUCLEAR WEAPONS Joseph M. Siracusa
NUMBERS Peter M. Higgins
NUTRITION David A. Bender
OBJECTIVITY Stephen Gaukroger
OCEANS Dorrik Stow
THE OLD TESTAMENT Michael D. Coogan
THE ORCHESTRA D. Kern Holoman
ORGANIC CHEMISTRY Graham Patrick
ORGANIZATIONS Mary Jo Hatch
PAGANISM Owen Davies
PAIN Rob Boddice
THE PALESTINIAN-ISRAELI CONFLICT Martin Bunton
PANDEMICS Christian W. McMillen
PARTICLE PHYSICS Frank Close
PAUL E. P. Sanders
PEACE Oliver P. Richmond
PENTECOSTALISM William K. Kay
PERCEPTION Brian Rogers
THE PERIODIC TABLE Eric R. Scerri
PHILOSOPHY Edward Craig
PHILOSOPHY IN THE ISLAMIC WORLD Peter Adamson
PHILOSOPHY OF LAW Raymond Wacks
PHILOSOPHY OF SCIENCE Samir Okasha
PHOTOGRAPHY Steve Edwards
PHYSICAL CHEMISTRY Peter Atkins
PILGRIMAGE Ian Reader
PLAGUE Paul Slack
PLANETS David A. Rothery
PLANTS Timothy Walker
PLATE TECTONICS Peter Molnar
PLATO Julia Annas
POLITICAL PHILOSOPHY David Miller
POLITICS Kenneth Minogue
POPULISM Cas Mudde and Cristóbal Rovira Kaltwasser
POSTCOLONIALISM Robert Young
POSTMODERNISM Christopher Butler
POSTSTRUCTURALISM Catherine Belsey
PREHISTORY Chris Gosden
PRESOCRATIC PHILOSOPHY Catherine Osborne
PRIVACY Raymond Wacks
PROBABILITY John Haigh
PROGRESSIVISM Walter Nugent
PROJECTS Andrew Davies
PROTESTANTISM Mark A. Noll
PSYCHIATRY Tom Burns
PSYCHOANALYSIS Daniel Pick
PSYCHOLOGY Gillian Butler and Freda McManus
PSYCHOTHERAPY Tom Burns and Eva Burns-Lundgren
PUBLIC ADMINISTRATION Stella Z. Theodoulou and Ravi K. Roy

PUBLIC HEALTH Virginia Berridge
PURITANISM Francis J. Bremer
THE QUAKERS Pink Dandelion
QUANTUM THEORY
John Polkinghorne
RACISM Ali Rattansi
RADIOACTIVITY Claudio Tuniz
RASTAFARI Ennis B. Edmonds
THE REAGAN REVOLUTION Gil Troy
REALITY Jan Westerhoff
THE REFORMATION Peter Marshall
RELATIVITY Russell Stannard
RELIGION IN AMERICA Timothy Beal
THE RENAISSANCE Jerry Brotton
RENAISSANCE ART
Geraldine A. Johnson
REVOLUTIONS Jack A. Goldstone
RHETORIC Richard Toye
RISK Baruch Fischhoff and John Kadvany
RITUAL Barry Stephenson
RIVERS Nick Middleton
ROBOTICS Alan Winfield
ROCKS Jan Zalasiewicz
ROMAN BRITAIN Peter Salway
THE ROMAN EMPIRE
Christopher Kelly
THE ROMAN REPUBLIC
David M. Gwynn
ROMANTICISM Michael Ferber
ROUSSEAU Robert Wokler
RUSSELL A. C. Grayling
RUSSIAN HISTORY Geoffrey Hosking
RUSSIAN LITERATURE Catriona Kelly
THE RUSSIAN REVOLUTION
S. A. Smith
SAVANNAS Peter A. Furley
SCHIZOPHRENIA Chris Frith and
Eve Johnstone
SCHOPENHAUER Christopher Janaway
SCIENCE AND RELIGION
Thomas Dixon
SCIENCE FICTION David Seed
THE SCIENTIFIC REVOLUTION
Lawrence M. Principe
SCOTLAND Rab Houston
SEXUALITY Véronique Mottier
SHAKESPEARE'S COMEDIES
Bart van Es
SHAKESPEARE'S SONNETS AND
POEMS Jonathan F. S. Post
SHAKESPEARE'S TRAGEDIES
Stanley Wells
SIKHISM Eleanor Nesbitt
THE SILK ROAD James A. Millward
SLANG Jonathon Green
SLEEP Steven W. Lockley and
Russell G. Foster
SOCIAL AND CULTURAL
ANTHROPOLOGY
John Monaghan and Peter Just
SOCIAL PSYCHOLOGY Richard J. Crisp
SOCIAL WORK Sally Holland and
Jonathan Scourfield
SOCIALISM Michael Newman
SOCIOLINGUISTICS John Edwards
SOCIOLOGY Steve Bruce
SOCRATES C. C. W. Taylor
SOUND Mike Goldsmith
THE SOVIET UNION Stephen Lovell
THE SPANISH CIVIL WAR
Helen Graham
SPANISH LITERATURE Jo Labanyi
SPINOZA Roger Scruton
SPIRITUALITY Philip Sheldrake
SPORT Mike Cronin
STARS Andrew King
STATISTICS David J. Hand
STEM CELLS Jonathan Slack
STRUCTURAL ENGINEERING
David Blockley
STUART BRITAIN John Morrill
SUPERCONDUCTIVITY
Stephen Blundell
SYMMETRY Ian Stewart
TAXATION Stephen Smith
TEETH Peter S. Ungar
TELESCOPES Geoff Cottrell
TERRORISM Charles Townshend
THEATRE Marvin Carlson
THEOLOGY David F. Ford
THINKING AND REASONING
Jonathan St B. T. Evans
THOMAS AQUINAS Fergus Kerr
THOUGHT Tim Bayne
TIBETAN BUDDHISM
Matthew T. Kapstein
TOCQUEVILLE Harvey C. Mansfield
TRAGEDY Adrian Poole
TRANSLATION Matthew Reynolds
THE TROJAN WAR Eric H. Cline

TRUST Katherine Hawley
THE TUDORS John Guy
TWENTIETH-CENTURY BRITAIN
Kenneth O. Morgan
THE UNITED NATIONS
Jussi M. Hanhimäki
THE U.S. CONGRESS Donald A. Ritchie
THE U.S. SUPREME COURT
Linda Greenhouse
UTILITARIANISM
Katarzyna de Lazari-Radek and
Peter Singer
UNIVERSITIES AND COLLEGES
David Palfreyman and Paul Temple
UTOPIANISM Lyman Tower Sargent
THE VIKINGS Julian Richards
VIRUSES Dorothy H. Crawford
VOLTAIRE Nicholas Cronk
WAR AND TECHNOLOGY Alex Roland
WATER John Finney
WEATHER Storm Dunlop
THE WELFARE STATE David Garland
WILLIAM SHAKESPEARE
Stanley Wells
WITCHCRAFT Malcolm Gaskill
WITTGENSTEIN A. C. Grayling
WORK Stephen Fineman
WORLD MUSIC Philip Bohlman
THE WORLD TRADE
ORGANIZATION Amrita Narlikar
WORLD WAR II Gerhard L. Weinberg
WRITING AND SCRIPT
Andrew Robinson
ZIONISM Michael Stanislawski

Available soon:

VETERINARY SCIENCE James Yeates
THE ENGLISH LANGUAGE
Simon Horobin
ORGANIZED CRIME
Georgios A. Antonopoulos and
Georgios Papanicolaou
THE PHILOSOPHY OF RELIGION
Tim Bayne
THE HISTORY OF PHYSICS
J. L. Heilbron

For more information visit our website

www.oup.com/vsi/

Warwick F. Vincent

LAKES

A Very Short Introduction

Great Clarendon Street, Oxford, OX2 6DP,
United Kingdom

Oxford University Press is a department of the University of Oxford.
It furthers the University's objective of excellence in research, scholarship,
and education by publishing worldwide. Oxford is a registered trade mark of
Oxford University Press in the UK and in certain other countries

© Warwick F. Vincent 2018

The moral rights of the author have been asserted

First edition published in 2018

Impression: 1

All rights reserved. No part of this publication may be reproduced, stored in
a retrieval system, or transmitted, in any form or by any means, without the
prior permission in writing of Oxford University Press, or as expressly permitted
by law, by licence or under terms agreed with the appropriate reprographics
rights organization. Enquiries concerning reproduction outside the scope of the
above should be sent to the Rights Department, Oxford University Press, at the
address above

You must not circulate this work in any other form
and you must impose this same condition on any acquirer

Published in the United States of America by Oxford University Press
198 Madison Avenue, New York, NY 10016, United States of America

British Library Cataloguing in Publication Data
Data available

Library of Congress Control Number: 2017954546

ISBN 978–0–19–876673–5

Printed in Great Britain by
Ashford Colour Press Ltd, Gosport, Hampshire

Links to third party websites are provided by Oxford in good faith and
for information only. Oxford disclaims any responsibility for the materials
contained in any third party website referenced in this work.

In memory of
Dennis A. Walter
(1938–2013)

Contents

Acknowledgements

The author would like to thank: François D. C. Forel for permission to include material from the publications of his great grandfather, François A. Forel; Amanda Toperoff for the superb graphics; Beatrix Beisner, Sylvia Bonilla, Rose Cory, Alexander Culley, George Kling, Michio Kumagai, Ulrich Lemmin, Isabelle Laurion, Connie Lovejoy, Sally MacIntyre, Stiig Markager, Frances Pick, Reinhard Pienitz, Milla Rautio, Geoffrey Schladow, Peter Vanrolleghem, and Adrien Vigneron for their valuable feedback on sections of the manuscript; Latha Menon and Jenny Nugee at Oxford University Press for their excellent editorial support; and the granting agencies that have financially supported the author's research on lakes, notably the Natural Sciences and Engineering Research Council of Canada and the Fonds de recherche du Québec—Nature et technologies.

List of illustrations

Chapter 1
Introduction

What is a lake? At first glance, this seems like such an easy question: a lake is simply a body of water surrounded by land. But this sterile, physical definition is only a beginning, and there are so many other more interesting ways to consider the nature and meaning of lakes. For freshwater biologists, a lake is an oasis in the landscape where microbes, plants, and animals form networks of interaction, and where species, food webs, and ecological processes await discovery. Many environmental scientists think of lakes in more chemical terms, as living reactors that exchange gases with the atmosphere. These are places that collect and transform materials washed in from the surrounding catchment, and where aquatic plants and algae produce new organic matter by photosynthesis. Some of my colleagues study microscopic fossils that occur in lake sediments, and for these researchers, lakes are rich storehouses of information that can tell us about the past, inform our present, and help guide our plans for the future.

For water engineers and society, lakes are essential resources to be managed, modified, even created, to address the ever increasing demand for drinking water, hydropower, fish production, and other ecosystem services. To maintain these services requires close attention to the balance of surface and groundwater inflows, evaporation, extractions, and outflows that together govern the amount of water remaining in the lake basin. Water is in terribly short supply in many parts of the world, and this balance of gains and losses is becoming ever more precarious and challenging to manage in our changing global climate.

In physical terms, a lake is a body of water that is constantly in motion, energized by the sun and the wind. Depending on season, the lake may be composed of layers that differ, sometimes surprisingly, in temperature, oxygen, colour, salt content, and many other properties, with periods of mixing each year that break down this layered structure. Lakes are connected, slow-moving conduits through the landscape: water moves from the inflows to the outflow of the lake, but this orderly, riverine flow path is continuously disrupted by wind-induced swirls, gyres, and counter-currents, while waves form and break, even out of view beneath the surface.

When I fly over the Canadian North to my field sites each summer, the landscape passes below as archipelagos of glittering freshwaters or, at more northerly, cooler latitudes, as snow-capped plates of lake ice set into the undulating tundra. In some of our work, we have been interested in the dispersal of microscopic life among these constellations of Arctic lakes and ponds, and how the individual waterbodies then select their ensemble of species. For those who study the evolution of fish and other aquatic life, the oldest lakes are island laboratories, where the processes of colonization, genetic shifts, and speciation can help us understand how the biodiversity of our planet has evolved, and how it is continuing to change. Charles Darwin even speculated about how life might have first arisen 'in some warm little pond with all sorts of ammonia and phosphoric salts'.

Lakes are the lowest points in the landscape, before they discharge ultimately (with some exceptions) to the ocean. In this way they can be thought of as integrators of their surroundings (Figure 1), reflecting the combined effects of water supply from their drainage basin (also called catchment or watershed), vegetation, geology, and the natural and human history of their environment. Lakes are

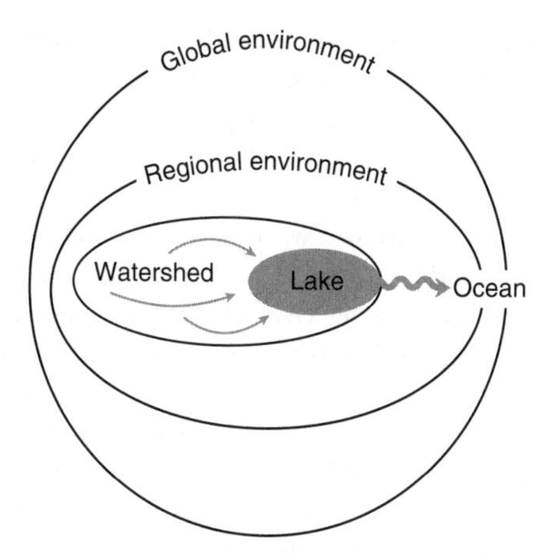

1. Lakes as sentinels, integrators, and conduits.

therefore indicators of environmental change, and can be viewed as sentinels of not only the current magnitude and legacy effects of local human activities, but also as regional and global sentinels of climate change, contaminant dispersal, and biodiversity shifts that are taking place throughout the world.

Then there is the question of size. Could Darwin's warm little pond be considered a lake? Some authors define ponds to be bodies of water of a depth that can be waded across, although in boggy wetlands, such a test would be ill advised. Another definition is that ponds freeze to the bottom while lakes do not, but this especially Canadian view of the aquatic world will not always be very helpful elsewhere; besides, ice-covered bottom waters are impressively resistant to freezing, even in Canada. Visitors to the English Lake District will find that the smaller bodies of water are called 'tarns', and larger ones are called 'lakes', 'meres', or 'waters', with no clear consensus on any of these terms. One of the most famous literary lakes in North America is called Walden Pond, and to further confuse this quagmire of terminology, the people of Newfoundland call most of their lakes 'ponds'—including Western Brook Pond that is 16 kilometres (km) long and 165 metres (m) deep. So all in all, it's best to consider lakes and ponds together, and to apply the word 'lakes' as a general term that encompasses the full range of waterbodies.

Size does become important if we want to know how many lakes there are in the world, in a particular country, or in our local surroundings. We need to set a cut-off value of minimum area, and with improvements in satellite remote sensing, and now increasingly with drones, the imaging threshold becomes smaller each year, as does the lower bound for lake inventories. High resolution satellites can readily detect lakes above 0.002 kilometres square (km^2) in area; that's equivalent to a circular waterbody some 50m across. Using this criterion, researchers estimate from satellite images that the world contains 117 million lakes, with a total surface area amounting to 5 million km^2. In Canada, we like to think that we have the largest number of lakes of any nation, including the North American Great Lakes that we share with the USA and that hold 20 per cent of the world's surface freshwater. But the world map on my office wall reminds me that Russia is also a large, lake-rich country, and that one of its lakes, Lake Baikal, is the deepest in the world and holds another 20 per cent of the surface freshwater on Earth.

The aim of this book is to provide a condensed overview of scientific knowledge about lakes, their functioning as ecosystems that we are part of and depend upon, and their responses to environmental change. Of course there are other, non-scientific, but equally varied reasons that lakes are important to human inquiry and culture. The mysterious, unsettling nature of deep or black waters has long held the interest of novelists and poets, for example Sylvia Plath in her 'Crossing the Water', and William Wordsworth's 'The Prelude', about his midnight traverse as a child across a dark troubling lake. The presence of mythical creatures that inhabit such depths, such as the Taniwha in Maori legends about New Zealand lakes and seas, has been the subject of oral histories in many parts in the world, and lakes may be of wider spiritual significance. In Bolivia and Peru, an ancient legend attributes the origin of the Incan civilization to the births of Manco Cápac and Mama Ocllo, brought up by the sun god Inti from the depths of Lake Titicaca. Lakes as mirrors and as multi-hued palettes of colour have captured the imaginations of artists, musicians, and writers of many cultures, and today attract multitudes of visitors to lakeshores each year. Some writing, including the classic haiku of Matsuo Basho, evokes the audible dimension of lakes and ponds ('the sound of water'), and in his monograph on dreams, Gaston Bachelard considers how the water of pools, ponds, fountains, lakes, and streams is a primary element of 'material imagination' and reverie.

This volume focuses on the science—and my hope is that this short introduction will allow the reader to view the next lake visited with a greater sense of wonder and a desire to learn more about the remarkable features that lie at and beneath its surface. Each chapter briefly introduces concepts about the physical, chemical, and biological nature of lakes, with emphasis on how these aspects are connected, the relationships with human needs and impacts, and the implications of our changing global environment. Lake science has a long history of observation and discovery, and many excellent textbooks are available that describe lake ecology in scholarly depth. These books encompass a broad sweep of established theory and new information, but the roots of much of this knowledge can be traced back to a career decision by a young scientist in the 19th century, at beautiful Lake Geneva, on the edge of the Swiss Alps.

Chapter 2
Deep waters

> I soon put to myself two options: create my own research laboratory in anatomy, histology and physiology, the subjects that I had to teach at the Faculty of Science... Or take for my laboratory and my aquarium this lake that offered me its mysteries and beckoned me to study them. My choice was soon made...
>
> F. A. Forel

Returning to the shores of Lake Geneva as a newly trained scientist and medical doctor, François A. Forel decided upon a career path that would guide his many decades of research, and that would ultimately lay the foundation for modern lake science. Forel was born and grew up on the Swiss side of the lake that lies on the border of Switzerland and France, and he was keenly aware of the wide-ranging importance of the lake to the people who lived around it. First and foremost, Lake Geneva was the primary source of drinking water for its lakeshore communities, including the rapidly growing city of Lausanne where Forel was appointed to teach at the Academy (today, the University of Lausanne).

Forel's father had captured his imagination as a child by bringing him out onto Lake Geneva, known in French as 'le Léman', to explore the ancient stilt-house villages found at several places in its inshore waters. The ruins of these Bronze Age settlements lay beneath the surface, and the archaeological artefacts found at these sites attested to the longstanding relationship between humans and the lake. Forel was also aware of the great commercial value of Lake Geneva for fisheries and transport, and he later quantified these resources in economic terms. He also appreciated the aesthetic appeal of the lake and its mountainous surroundings, and he enjoyed the company of landscape artist François L. D. Bocion. But he was especially taken by the idea that Lake Geneva's deep, blue waters held scientific mysteries and secrets, many of which could eventually be revealed by careful study.

Forel had returned to Lausanne and the family home in nearby Morges in 1867, at the age of 26, after some eleven years of schooling and medical training in Switzerland, France, and Germany. His decision to study 'all aspects of the lake' was initially greeted with some concern by his former professor in Germany, who advised the young man to adopt a more focused and specialized approach. Not to be dissuaded, Forel launched his research on a myriad subjects, from waves and currents, the penetration of sunlight, and the nature of chemicals in the water, to studies on the plants, animals, and microbes that live throughout the lake. It was not until many years later, however, that he brought all of these disparate studies together into an overall synthesis.

In the preface to volume one of his comprehensive monograph on Lake Geneva, published in 1892, Forel coined the word 'limnology', from the Greek 'limne' for lake, and he defined this new integrative science as 'the oceanography of lakes'. Today, limnology has been extended to include rivers, wetlands, and even estuaries, but its primary focus remains lakes and ponds. There are many professional societies around the world for researchers in this field, notably the International Society of Limnology (SIL) and the Association for the Sciences of Limnology and Oceanography (ASLO). The word limnology, however, is a scientific term that is not well known outside the field. The word 'lake' that we use in English for bodies of freshwater is derived not from a familiar root (like Greek 'okeanos' in oceanography) but from the Latin word 'lacus' meaning basin. On the other hand, the concept of limnology is intuitively straightforward and appealing, and is highly relevant to our present-day goals of lake protection, restoration, and management.

For Forel, the science of lakes could be subdivided into different disciplines and subjects, all of which continue to occupy the attention of freshwater scientists today (Figure 2). First, the physical environment of a lake includes its geological origins and setting, the water balance and exchange of heat with the atmosphere, as well as the penetration of light, the changes in temperature with depth, and the waves, currents, and mixing processes that collectively determine the movement of water. Second, the chemical environment is important because lake waters contain a great variety of dissolved materials ('solutes') and particles that play essential roles in the functioning of the ecosystem. Third, the biological features of a lake include not only the individual species of plants, microbes, and animals, but also

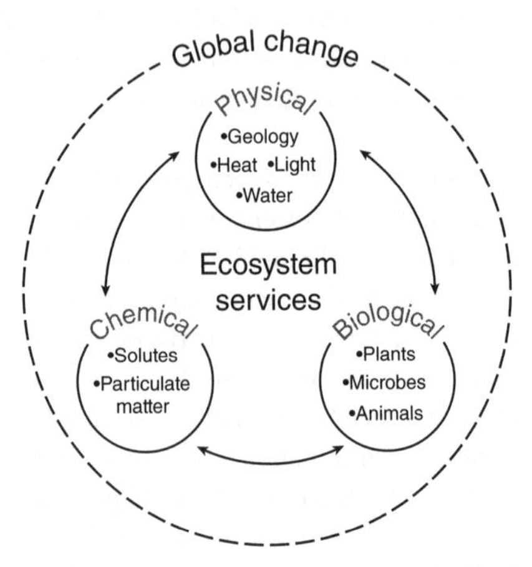

2. Interactions that affect lake ecosystem services in the changing global environment.

their organization into food webs, and the distribution and functioning of these communities across the bottom of the lake and in the overlying water.

There were two aspects of Forel's new science that set his thinking apart from many of his contemporaries, and even from that of many specialists subsequently. As he stated in his earliest definition of limnology, studies of a lake from different points of view can inform each other and show connections, producing an integrated picture of the ecosystem. The interactions among the physical, chemical, and biological properties of the lake were of special interest. For example, he showed how terrestrial organic materials could give rise to the greenish hues of the lake and affect the clarity of the water, how the weathering of rocks in the surrounding catchment affected the mineral chemistry of the lake, and how bottom-dwelling animals were connected to the life and death of plankton in the overlying waters. In contrast to the popular idea of a lake as an enclosed microcosm, he wrote in 1891 that:

> Rather, it communicates with the rest of the world, via its constant exchange of gases with the overlying atmosphere, via its outflow of water carrying away dissolved and non-dissolved substances, and via its tributaries that deliver new materials into the lake.

The second aspect concerned human beings. Forel recognized that lakeshore residents were actually part of the Lake Geneva ecosystem, and that they depended on the lake in many ways to provide them with services such as safe drinking water, the commercial fishery, a waterway for transport of people and cargo, and the aesthetic pleasure and mental well-being of living close to water. All of these benefits can become less available as a result of human impacts on the environment, and Forel observed many examples in Lake Geneva, including the defective regulation of water level, the introduction of invasive aquatic species, and contamination of the lake with human pathogens from sewage. This idea that people are a component part of the ecosystem is a concept that was not fully recognized through much of the 20th century. Yet it is crucially important today as we confront the increasing impacts of global change, and the challenges of sustaining the biosphere that we are both part of and dependent upon.

Birth and death of lakes

Forel devoted many pages of the first of his three-volume treatise on Lake Geneva to considering the possible origins of its basin, and he also described the way that sediments were accumulating in the lake, especially those derived from the glacier-fed, upper Rhône River with its milky suspension of mineral particles. This continuous accumulation of materials on the lake floor, both from inflows and from the production of organic matter within the lake, means that lakes are ephemeral features of the landscape, and from the moment of their creation onwards, they begin to fill in and gradually disappear. The world's deepest and most ancient freshwater ecosystem, Lake Baikal in Russia (Siberia), is a compelling example: it has a maximum depth of 1,642m, but its waters overlie a much deeper basin that over the twenty-five million years of its geological history has become filled with some 7,000m of sediments.

Lakes are created in a great variety of ways: tectonic basins formed by movements in the Earth's crust, the scouring and residual ice effects of glaciers, as well as fluvial, volcanic, riverine, meteorite impacts, and many other processes, including human construction of ponds and reservoirs. Tectonic basins may result from a single fault, as in Lake Baikal and Lake Tanganyika (eastern Africa), or from a series of intersecting fault lines. Lake Tahoe (USA), for example, was created by block faulting giving rise to its rectangular, horse-trough shaped basin with a great average (300m) as well as maximum (501m) depth. The oldest and deepest lakes in the world are generally of tectonic origin, and their persistence through time has allowed the evolution of endemic plants and animals; that is, species that are found only at those sites.

Some of the best known examples of tectonic lakes and their endemic fauna are in the African Rift Valley, where each basin has been isolated for a sufficiently long time to allow adaptive radiation of numerous fish species. Lake Malawi has the largest number, with more than 850 fish. Most of these are endemic and are distributed among eleven families, with cichlids as the most species-rich. In Lake Tanganyika, sixteen families of fish are represented, including 200 cichlids, while the vast waters of Lake Victoria (68,800km^2; maximum depth 84m) are thought to have been the habitat for more than 500 fish species in the past. These lakes are facing pressure from agricultural development, fisheries, and new species introductions, for example of the Nile perch into Lake Victoria, and this combination of increased predation, competition, and changing water quality has resulted in the loss of many endemic species, perhaps as many as 200 in Lake Victoria.

Other ancient tectonic lakes with endemic species include Lake Biwa, Japan, with seventeen endemic fish, for example the giant Lake Biwa catfish *Silurus biwaensis*; Lake Titicaca, whose endemic fauna includes fifteen species of pupfish (*Orestias*) and a giant water frog (*Telmatobius culeus*); Lake Ohrid, with endemic sponges and fifty snail species; and Lake Baikal, which is the habitat for more than 1,000 endemic species including phytoplankton such as the diatom *Aulacoseira baicalensis*, invertebrates including amphipods and the dominant zooplankton species *Epischura baikalensis*, fish including the Baikal sculpin, and the sole exclusively freshwater species of seal, *Pusa sibirica*.

In terms of total numbers, most of the world's lakes (including those of the English Lake District; Figure 3) owe their origins to glaciers that during the last ice age gouged out basins in the rock and deepened river valleys. These lakes include the deep lakes of Europe such as Lake Geneva (310m deep), Lake Constance

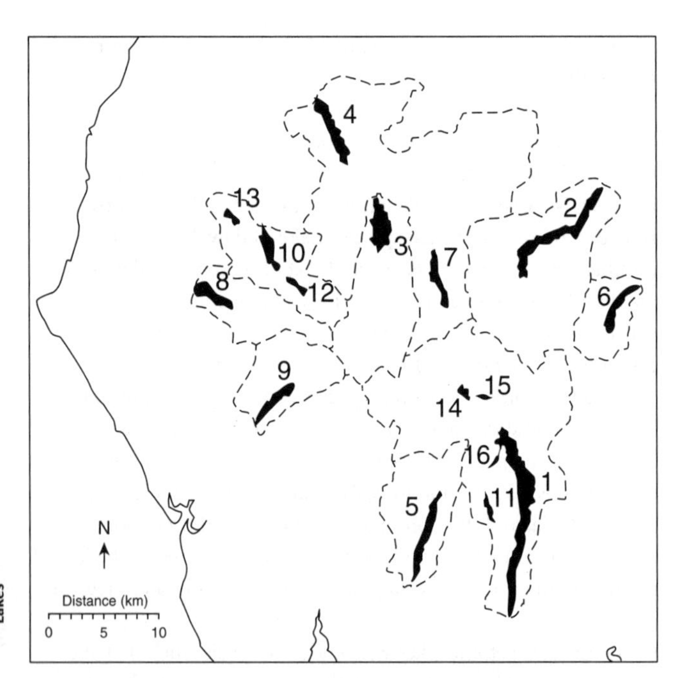

3. Lakes of the English Lake District and their catchments. Situated in the northwest corner of England, the English Lake District (see Table 1) has been a major centre for freshwater studies from the 1920s onwards through the Freshwater Biological Association of the United Kingdom (FBA), and more recently through the Centre for Ecology & Hydrology (CEH). The lakes are arranged like the spokes of a wheel; this pattern is thought to be derived from a radial pattern of drainage over a central dome, which was then eroded and the valleys deepened by Pleistocene glaciers.

(251m), Lago Maggiore (372m), and Lake Como (425m); the freshwater lochs of Scotland including Loch Ness (227m) and Loch Morar (310m); the North American Great Lakes including Lake Michigan (281m) and Lake Superior (406m); and lakes in the South Island of New Zealand such as Lake Wakatipu (380m) and Lake Hauroko (462m). These glacial processes have also scratched out numerous shallower depressions in the ground,

Table 1 Lakes of the English Lake District (numbers relate to Figure 3)

Number	Lake	Area (km^2)	Maximum depth (m)	Mean depth (m)
1	Windermere	14.8	64.0	21.3
2	Ullswater	8.9	62.5	25.3
3	Derwent Water	5.3	22.0	5.5
4	Bassenthwaite Lake	5.3	19.0	5.3
5	Coniston Water	4.9	56.0	24.1
6	Haweswater	3.9	57.0	23.4
7	Thirlmere	3.3	46.0	16.1
8	Ennerdale Water	3.0	42.0	17.8
9	Wastwater	2.9	76.0	39.7
10	Crummock Water	2.5	44.0	26.7
11	Esthwaite Water	1.0	15.5	6.4
12	Buttermere	2.0	28.6	16.6
13	Loweswater	0.6	16.0	8.4
14	Grasmere	0.6	21.6	7.7
15	Rydal Water	0.3	20.0	7.0
16	Blelham Tarn	0.1	14.5	6.8

for example on the Precambrian granites of northern Canada. Here the rocky landscape is replete with lakes that are just a few thousand years old, and the bottoms of many of these young Arctic waterbodies are coated with only the thinnest layer of lake sediments.

As the glaciers retreated, their terminal moraines (accumulations of gravel and sediments) created dams in the landscape, raising water levels or producing new lakes. These moraine-dammed, glaciated basins include the beautiful lakes of the lake district of southern Chile, such as Lake Llanquihue (317m; also affected by volcanic activity) and Riñihue Lake (323m). One of the largest (1,850km^2), deepest (586m), moraine-dammed lakes in South America is known by two names, since it extends across Patagonia from Argentina, where it is called Lake Buenos Aires, to Chile, where it is named General Carrera Lake. During glacial retreat in many areas of the world, large blocks of glacial ice broke off and were left behind in the moraines. These subsequently melted out to produce basins that filled with water, called 'kettle' or 'pothole' lakes. Such waterbodies are well known across the plains of North America and Eurasia.

As glaciers and expanding ice caps bulldoze their way across the landscape, their terminus may end in a basin of meltwater that is pushed ahead by the glacial ice flow. When the glaciers melt and retreat, these 'proglacial lakes' may expand dramatically until their ice dams dwindle in size or are breached. One of the most spectacular examples in North America is the twin set of proglacial lakes, Agassiz and Ojibway, that formed in front of the Laurentian Ice Sheet during the last Ice Age. At its maximum some 13,000 years ago, Lake Agassiz extended over an estimated 440,000km^2, almost twice the total area of the North American Great Lakes today. When a section of the ice cap finally collapsed in northern Hudson Bay around 8,200 years ago, the water gushed out and the merged Lake Agassiz-Ojibway drained almost completely, raising global sea levels by 0.8m or more. This catastrophic event is thought to have caused abrupt changes in oceanic circulation and climate, which in turn may have altered the human migration patterns and agriculture of prehistoric cultures in Europe.

The most violent of lake births are the result of volcanos. The craters left behind after a volcanic eruption can fill with water to form small, often circular-shaped and acidic lakes. The highest lake in the world is a small waterbody at 6,390m that occupies the crater of Nevado Ojos del Salado, an active volcano on the border of Chile and Argentina. Much larger lakes are formed by the collapse of a magma chamber after eruption to produce caldera lakes. The largest known of such lake-forming events was the supervolcanic eruption that formed Lake Taupo (at present 616km^2; depth 186m) in the central North Island of New Zealand, 26,500 years ago. This ejected more than 1,000 kilometres cubed (km^3) of material, and the resultant collapse created a huge caldera that filled with water. This lake has also undergone subsequent eruptions, most recently about 5,000 years ago, and today there are geothermal features within and near the lake attesting to the geologically active nature of the area.

Craters formed by meteorite impacts also provide basins for lakes, and have proved to be of great scientific as well as human interest. One of the most distinctive is Lake Manicouagan, a large (1,942km^2; depth 350m), ring-shaped lake in central Quebec caused by the impact of a 5km diameter asteroid some 214 million years ago. Much further north in the sub-Arctic region of Quebec, Canada, lies Pingualuk Lake, almost perfectly circular in shape with a diameter of 3km (Figure 4). This lake has long been known to the Inuit, who consider its crystalline waters to be imbued with healing powers and call it the 'Crystal Eye of Nunavik'. The crater it occupies is the result of a meteorite impact 1.4 million years ago, and because of its considerable depth (400m, with a current lake depth of 267m), the waterbody likely remained unfrozen beneath the thick ice sheet during glacial periods as a subglacial lake, perhaps with conditions similar to those found today in the subglacial lakes Vostok, Whillans, and Ellsworth in Antarctica. The deep sediments of Pingualuk Lake have been sampled by lake scientists to provide an uninterrupted record of past climates over several glacial–interglacial cycles.

There was a time when limnologists paid little attention to small lakes and ponds, but, this has changed with the realization that

4. Pingualuk Lake, the 'Crystal Eye' of northern Quebec.

although such waterbodies are modest in size, they are extremely abundant throughout the world and make up a large total surface area. Furthermore, these smaller waterbodies often have high rates of chemical activity such as greenhouse gas production and nutrient cycling, and they are major habitats for diverse plants and animals, including water fowl. A prominent example is Arctic thaw ponds and lakes ('thermokarst lakes') that are formed by the thawing of ice-rich, perennially frozen ground ('permafrost'). These occur in massive abundance over many parts of the northern landscape, with a total collective area of more than 250,000km^2. These waterbodies are undergoing rapid changes in response to permafrost thaw and degradation because of global climate change. In some areas they are disappearing as a result of evaporation, infilling, or drainage, while in other areas they are expanding in size and abundance. These waterbodies are hotspots of microbial activity that break down previously frozen stores of ancient soil carbon into carbon dioxide and methane, which are then released to the atmosphere.

The underwater shape of lakes

A bathymetric map showing the three-dimensional form or 'morphometry' of the basin is an essential starting point for any lake study. Such maps are increasingly available in digital format, yet there are still parts of the world where this basic information is completely lacking. Once a bathymetric map is obtained, several important values can be calculated. The first step is to calculate the area within each depth contour, which is most easily done with a geographic information system (GIS) software package. These areas can then be plotted on a graph as a function of depth. This area-depth plot is called a 'hypsographic curve', and it allows some useful statistics about the lake to be easily determined.

With the illustrative curve for Lake Baikal (Figure 5), we can ask the question: how much of this ancient lake has a depth greater than 500m? For most lakes of the world, the answer would be

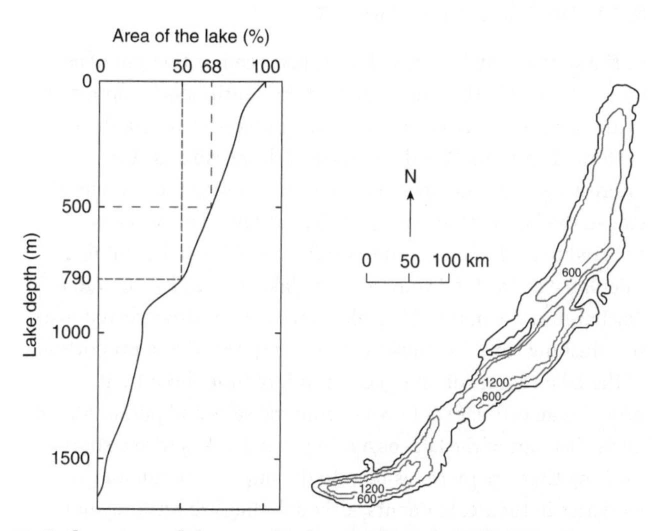

5. Bathymetry and the area-depth curve for Lake Baikal, Russia.

none, since even their maximum depth is much shallower. However, the Lake Baikal bathymetric map shows that it contains three deep basins, and the area-depth graph, which turns the complicated morphometry of the lake into a single curve, reveals that an impressive 68 per cent of the lake is deeper than 500m. Similarly, the curve can be easily read off to show that 50 per cent of the lake is 790m or deeper. The different depth layers in the hypsographic curve can be summed to calculate the total volume of the lake, which for Lake Baikal is 23,000km^3, equivalent to flooding all of England with freshwater to a depth of 176m. The total volume can then be divided by the area of the lake to give another limnological measure, the mean lake depth; for Lake Baikal this is 744m. In general, lakes with a larger mean depth tend to be of greater transparency and higher water quality, but human impacts on such lakes can cause a severe deterioration of these features, as has even been observed in the waters of Lake Baikal.

Rise and fall of water levels

In the simplest hydrological terms, lakes can be thought of as tanks of water in the landscape that are continuously topped up by their inflowing rivers, while spilling excess water via their outflow (Figure 6). Based on this model, we can pose the interesting question: how long does the average water molecule stay in the lake before leaving at the outflow? This value is referred to as the water residence time, and it can be simply calculated as the total volume of the lake divided by the water discharge at the outlet. This lake parameter is also referred to as the 'flushing time' (or 'flushing rate', if expressed as a proportion of the lake volume discharged per unit of time) because it provides an estimate of how fast mineral salts and pollutants can be flushed out of the lake basin. In general, lakes with a short flushing time are more resilient to the impacts of human activities in their catchments, although they are certainly not immune to such effects.

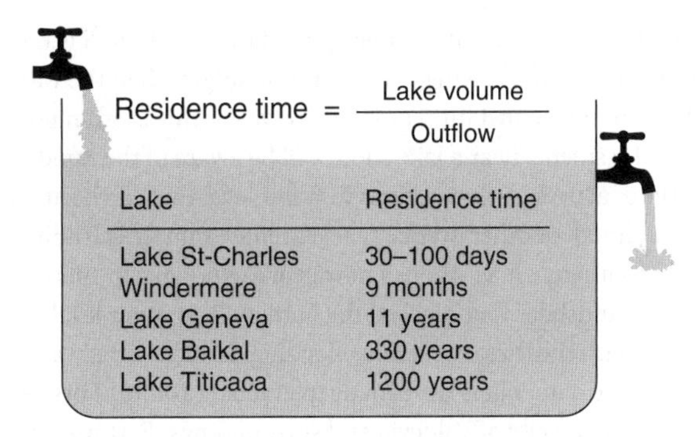

Residence time = Lake volume / Outflow
Lake
Residence time
Lake St-Charles
30–100 days
Windermere
9 months
Lake Geneva
11 years
Lake Baikal
330 years
Lake Titicaca
1200 years

6. Differences in water residence time among lakes.

Each lake has its own particular combination of catchment size, volume, and climate, and this translates into a water residence time that varies enormously among lakes (Figure 6). Lake St-Charles, for example, is our drinking water reservoir in Quebec City, and is a dammed river basin that is fed by a large catchment ($169km^2$) relative to the size of the lake ($3.6km^2$). The residence time for this reservoir is therefore only one to a few months, depending on the time of year. At the other extreme, Lake Titicaca has a small discharge relative to its volume, and its calculated residence time is more than 1,000 years.

A more accurate approach towards calculating the water residence time is to consider the question: if the lake were to be pumped dry, how long would it take to fill it up again? For most lakes, this will give a similar value to the outflow calculation, but for lakes where evaporation is a major part of the water balance, the residence time will be much shorter. This is the case for Lake Titicaca, where the true water residence time is only eighty years based on inputs (not the 1,200 years based on outflow, as in Figure 6), because more of its water is lost by evaporation than through the outflow. This lack of complete flushing also means that salts are concentrated in the lake, and the water is slightly brackish (salinity around 0.7 parts per thousand (ppt)).

If the total volume of water entering the tank in Figure 6 is the same as the volume leaving via its outflow spigot, then the water level will remain constant. For lakes, however, this is often not the case, and residents near a lake shore will be aware of the rapid, sometimes alarming, fluctuations in water level that can occur after a heavy rainfall or other events. One extreme example is Riñihue Lake in southern Chile, where a powerful earthquake in 1960 triggered landslides that blocked the outflow. The water level rose by 20m. With the threat of serious flooding damage if the waters breached the dam, plans were made to evacuate 100,000 people in the downstream city of Valdevia and surroundings. Fortunately, much of this water could be released in a controlled fashion by excavation of release channels over the subsequent weeks.

Large variations can also occur as a result of the natural cycles of river flow. The most pronounced example is the Amazon River and its extensive floodplain, known as the 'varzéa'. Many fish depend upon the annual flooding cycle, which allows them to swim into the inundated forest to feed on terrestrial insects, spiders, nuts, seeds, and flowers. One of the numerous lakes is Lago Calado, which lies 60km upstream of Manaus, the Brazilian city at the confluence of the Rio Solimões and the Rio Negro, in the heart of the Amazonian rainforest. Like other lakes of the region, it has a complex, dendritic shape, and when its basin floods with the cappuccino-coloured waters of the Rio Solimões, the lake level rises by 10m, and its area expands by a factor of four. Varzéa lakes contain floating meadows of the grasses *Paspalum repens* and *Echinochloa polystachya*, and these rise and fall with the seasons, providing lush island habitats for insects, birds, and snakes.

Climate change has a major effect on lake water levels by altering the balance of inflows and evaporative losses. One of the most striking examples is Lake Chad, at the edge of the Saharan Desert. Given its shallow depth (maximum depth 11m; mean depth 1.5m), it is highly sensitive to variations in rainfall, both with season and in the longer term. Over the last fifty years, there has been a

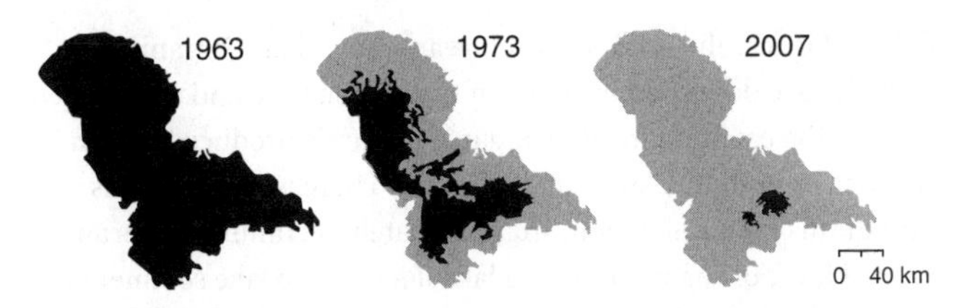

7. The shrinking of Lake Chad in central Africa caused by ongoing dry conditions.

massive contraction in lake area (Figure 7) associated with a drier climate, compounded by inefficient damming, irrigation, and land clearance for agriculture. Violent conflicts have emerged between fishermen and farmers, who have opposing needs for the water. The geological record shows that Lake Chad has undergone profound changes in the past, from palaeolake Mega-Chad of more than 1 million km^2, to periods of almost complete dryness. The full loss of this resource today would have devastating consequences for more than thirty million people who currently depend on water from the lake.

Drops in lake level can also reveal some surprises. Lake Kinneret (area 167km^2; maximum depth 43m) in Israel is also known as the Sea of Galilee, a place that figures prominently in the New Testament of the Bible. During a period of drought in the late 1980s, the water level of this freshwater lake dropped 9m and revealed the presence of a stone-age settlement of oval huts (an archaeological site now called Ohalo II), dating back 23,000 years before the present. These are among the oldest known dwellings in the world, and are evidence of humankind's longstanding association with lakes.

Lake sediments as archives

Lake Chad is an extreme example of how lakes respond to environmental change, but even subtle variations in climate and human activities are registered in lakes, and in a way that can be decoded by careful analysis. Each year, mineral and organic particles are deposited by wind on the lake surface and are washed in from the catchment, while organic matter is produced within the lake by aquatic plants and plankton. There is a continuous rain of this material downwards, ultimately accumulating as an annual layer of sediment on the lake floor. These lake sediments are storehouses of information about past changes in the surrounding catchment, and they provide a long-term memory of how the limnology of a lake has responded to those changes. The analysis of these natural archives is called 'palaeolimnology' (or 'palaeoceanography' for marine studies), and this branch of the aquatic sciences has yielded enormous insights into how lakes change through time, including the onset, effects, and abatement of pollution; changes in vegetation both within and outside the lake; and alterations in regional and global climate.

The sampling approach in palaeolimnology begins with the coring of the lake sediments. A variety of portable devices have been developed to obtain short cores for analysis of the last few hundred years of the sediment record, while much heavier drilling equipment is used to core the sediment to much greater depth, and to extend the record back to tens of thousands of years or longer. For example, at Lake El'gygytgyn, a crater lake in Siberia formed by a meteorite impact 3.58 million years ago, a 400m long sediment and rock sequence was obtained to capture a continuous 3.6 million year record of Arctic climate change, including the Pliocene–Pleistocene transition. At Lake Biwa, Japan, formed by tectonic processes three million years ago, the upper 250m of lake sediments from a 1.4km long core provided records that extend back to 430,000 years before the present.

Sampling for palaeolimnological analysis is typically undertaken in the deepest waters to provide a more integrated and complete picture of the lake basin history. This is also usually the part of the lake where sediment accumulation had been greatest, and where the disrupting activities of bottom-dwelling animals (‘bioturbation’ of the sediments) may be reduced or absent. Once the core is brought up to the surface, the sediment is extruded through its core barrel and split into sections. The age of the upper layers is established by measuring the radioisotopes cesium-137 and lead-210, which provide a chronology of sediment dates for the last 150 years or so. For deeper, older layers, the dates are established by analysis of the radioisotope carbon-14. With several of these radioisotopic analyses down the core, dates can then be estimated by interpolation for each and every layer, and these time-stamped strata then analysed for evidence of change.

At first inspection under the microscope, a sample from one of those core slices will seem little more than a smear of nondescript particles that vary randomly in shape and size. However, with careful observation, the undecomposed remains of all manner of terrestrial and aquatic life can be discerned among the particles, and in many cases identified to their species of origin. These ‘microfossils’ include pollen grains that have been blown or washed into the lake, and that have resisted decomposition in the sediments because of their hardy outer walls. The distinctive shape of the pollen grains allows them to be identified to genus and even species, and the lake sediments thereby provide a record of changes in plant community structure in the surrounding landscape.

Some of the most informative microfossils to be found in lake sediments are diatoms, an algal group that has cell walls (‘frustules’) made of silica glass that resist decomposition. Each lake typically contains dozens to hundreds of different diatom species, each with its own characteristic set of environmental preferences as well as distinctive cell wall shape and ornamentation that can be used to identify them from their microfossil remains in the sediments. A widely adopted approach is to sample many lakes and establish a statistical relationship or ‘transfer function’ between diatom species composition (often by analysis of surface sediments) and a lake water variable such as temperature, pH, phosphorus, or dissolved organic carbon. This quantitative species–environment relationship can then be applied to the fossilized diatom species assemblage in each stratum of a sediment core from a lake in the same region, and in this way the physical and chemical fluctuations that the lake has experienced in the past can be reconstructed or 'hindcast' year-by-year. Other fossil indicators of past environmental change include algal pigments, DNA of algae and bacteria including toxic bloom species, and the remains of aquatic animals such as ostracods, cladocerans, and larval insects.

One example of tapping into these historical archives is a study at Walden Pond (Figure 8), a kettle lake near the city of Boston in Massachusetts, USA, with a maximum depth of about 30m and surface area of 25 hectares (ha). This is a site well known to readers of American literature, because it was here that naturalist, essayist, philosopher, and historian Henry David Thoreau spent two years on its shores, from 4 July 1845 to 6 September 1847. The experience formed the basis of his classic work *Walden*, published in 1854, and it inspired his meditation on the natural world, writing how 'A lake is a landscape's most beautiful and expressive feature. It is earth's eye; looking into which the beholder measures the depth of his own nature'.

Thoreau kept detailed records in his daily journal of many features of the lake, including notes on the layering of the water,

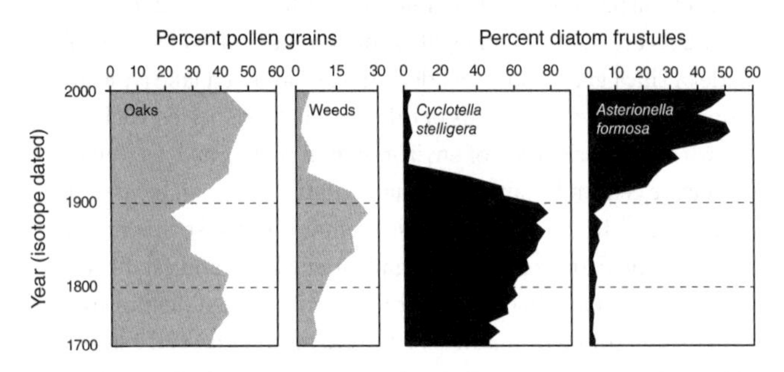

8. Environmental changes over the last 300 years recorded in the lake sediments of Walden Pond, USA.

with warm above and cold below. This was some twenty years before Forel began his studies at Lake Geneva, and half a century before James G. Needham, Professor of Limnology at Cornell University, wrote the first English textbook on freshwater ecology. It has been suggested that Thoreau, despite his reluctance to embrace all aspects of science, might well be considered North America's first limnologist.

The pollen profiles in the 28-centimetres (cm) long sediment core shown in Figure 8 bear witness to the clearing of forest by the early settlers of New England, with a reduction in oaks and the expansion of agricultural lands, hence an increase in weed pollen. It is ironic that while Thoreau was making his observations about the spiritual value of lakes and forests, this wholesale destruction of forests was at its peak, with 80 per cent of the land ultimately clear-felled and converted to agriculture. Towards the beginning of the 20th century, a reversal of this process is apparent in the pollen records, as rural populations left their homes to work in cities and allowed the uneconomic farmlands to revert back to forest, as indicated in the sediment records by a reduction in weed pollen and a proportionate increase in oak pollen.

The diatom analyses show that other changes were in motion at Walden Pond in the period 1880 to the present (Figure 8). Certain species such as *Cyclotella stelligera* rose gradually in prominence, and then suddenly gave way to species such as *Asterionella formosa* that are more characteristic of nutrient-enriched waters. The application of a transfer function for phosphorus to the full assemblage of diatoms in the sediments indicated that this key nutrient stimulating algal blooms rose abruptly in concentration from the 1920s onwards, associated with the recreational development of the lake. Thoreau's celebration of nature while living alone at Walden Pond is now shared by thousands of visitors each summer, and ongoing careful management will be required to preserve the cultural and ecological values of this iconic lake for future generations.

Chapter 3
Sunlight and motion

> When the waves are lively . . . the proportion of blue mingles with the reflected colours, and varies with the shape and size and direction of the waves.
>
> F. A. Forel

François Forel devoted much of his treatise on Lake Geneva to the physical environment of the lake: light, temperature, wind, waves, currents, and mixing. He noticed that water at times moved in and out through a narrow opening to the harbour near his house at Morges with surprising speed and regularity, and he realized that this had something to do with a see-saw rocking motion that extended across the entire lake. He talked with fishermen, who told him strange stories about how their nets set deeply beneath the surface were dragged by currents, but in the opposite direction to that of the wind. He realized that the lake was not simply a basin of well-mixed water, like an aerated fish tank, but instead was composed of layers with different temperatures, and that this layering changed according to season.

Forel was especially fascinated by the interplay between sunlight and water, an interest encouraged by watching his friend the painter François Bocion at work as he captured Lake Geneva's multi-coloured hues of sky, clouds, and water on the canvas. Forel noted changes in the turbidity of inshore waters, from clear to opalescent to muddy, and he surmised that water transparency could be a simple yet powerful indicator of the state of health of a lake ecosystem. Freshwater scientists today are very much aware of the importance of all of these features, which define the physical habitat characteristics of lakes and strongly influence their chemistry, biology, and ecosystem services (Figure 2).

Clear or murky waters

Soon after Forel began his studies at Lake Geneva, he learned of a simple method to measure water transparency, and he became the first person to apply it to lakes and develop a standardized protocol. The method had been conceived for the clear blue waters of the Mediterranean Sea by a priest and scientific advisor to the Pope, Pietro Angelo Secchi. Working on board the Papal Navy ship 'Immacolata Concezione', Secchi undertook studies to better understand the interaction between sunlight and the sea. His elegant approach was to simply lower a white disc and note the depth at which it could no longer be seen.

Forel seized upon this idea from Secchi and he established a formal protocol: take a 20cm diameter white disk, note the depth where the Secchi disk disappears, bring it up again slowly noting the depth at which it reappears, and calculate the 'Secchi depth' as the average of those two values. Secchi had used disks of different colours and sizes, including one that was 2.37m in diameter. Forel recommended 20cm because of its portability for travelling, and also because he saw little difference with a larger, 35cm version. He employed both a white painted, zinc metal disk and a white glazed, ceramic dinner plate, and noted that while the former was less fragile, the latter retained its white coloration for longer. These days, 20cm or 30cm diameter metal disks are routinely used in lake studies, and are usually painted with alternating quadrants of black and white to enhance contrast.

Values of Secchi depth range from a few centimetres in highly polluted waters with algal blooms, to tens of metres in the world's most transparent lakes. The overall record is held by the Weddell Sea, Antarctica, where a 20cm diameter Secchi disk could be seen to a depth of 79m; this is close to the theoretical limit for visibility in pure water. For lake waters, the record is Crater Lake, Oregon, USA, where a 1m diameter Secchi disk (somewhat outside Forel's specifications, but fine for Father Secchi) could be seen to 44m.

In lake and ocean studies, the penetration of sunlight into the water can be more precisely measured with an underwater light meter (submersible radiometer), and such measurements always show that the decline with depth follows a sharp curve rather than a straight line (Figure 9). This is because the fate of sunlight streaming downwards in water is dictated by the probability of the photons being absorbed or deflected out of the light path; for example, a 50 per cent probability of photons being lost from the light beam by these processes per metre depth in a lake would result in sunlight values dropping from 100 per cent at the surface to 50 per cent at 1m, 25 per cent at 2m, 12.5 per cent at 3m, and so on. The resulting exponential curve means that for all but the clearest of lakes, there is only enough solar energy for plants, including photosynthetic cells in the plankton (phytoplankton), in the upper part of the water column.

The depth limit for underwater photosynthesis or primary production is known as the 'compensation depth'. This is the depth at which carbon fixed by photosynthesis exactly balances the carbon lost by cellular respiration, so the overall production of new biomass (net primary production) is zero. This depth often corresponds to an underwater light level of 1 per cent of the sunlight just beneath the water surface (Figure 9). The production of biomass by photosynthesis takes place at all depths above this level, and this zone is referred to as the 'photic zone'. At deeper,

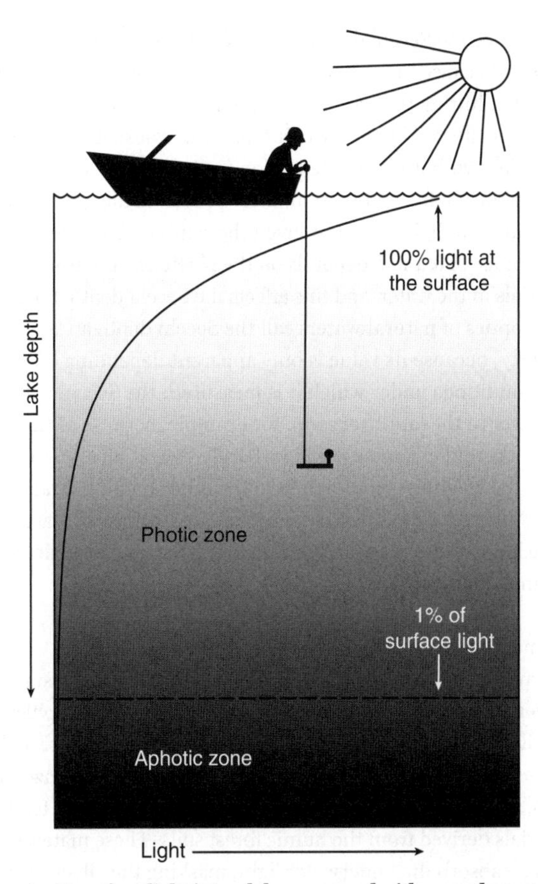

9. Penetration of sunlight into a lake measured with an underwater light meter.

less illuminated depths, no photosynthetic growth is possible and biological processes in this 'aphotic zone' are mostly limited to feeding and decomposition.

A Secchi disk measurement can be used as a rough guide to the extent of the photic zone: in general, the 1 per cent light level is about twice the Secchi depth. However, this is not always accurate because the stream of photons passing down through the water, hitting the Secchi disc, and returning to our eyes at the surface is affected by the two different processes: absorption (given the symbol '*a*') and deflection (called 'scattering' and given the symbol '*b*'). In combination, these processes dictate the overall attenuation of light, which is given the symbol '*c*'. The relative importance of *a* and *b* depends on the particles and dissolved materials in the water, and this affects the Secchi depth. Specialists in the optics of natural waters call the Secchi depth an 'apparent property', because its value is only apparent depending on the light conditions under which it is measured; the Secchi will be shallower in the late afternoon, for example, because of the lower sun angle, and of course it will be barely seen at all at night, even under the brightest moonlight. On the other hand, *a*, *b*, and *c* are called 'inherent properties' because they are intrinsic characteristics of the lake water that are not affected by sunlight conditions at the time of measurement.

In general, more algal particles in the water cause more absorption and scattering, and the Secchi depth is a measure of nutrient pollution and algal enrichment of a lake ('eutrophication', described in Chapter 7). However, for lakes in heavily forested regions such as the circumpolar boreal zone and the Amazonian basin, the inflowing waters are highly charged in brown, tea-like materials derived from the humic forest soils. These materials strongly absorb the underwater light, masking the effects of algae. At the opposite extreme are waters with lots of reflective mineral particles in suspension. These waters have high *b* values, and many of the photons reflected by the Secchi disc are scattered out of the light beam that comes back to our eyes, but they are still available for photosynthesis in the lake. This means that the Secchi depth has to be multiplied by a larger number (up to three in some waters laden with mineral particles) to estimate the extent of the photic zone.

Despite its limitations, the Secchi disk has proved to be an extremely valuable tool for lake studies and science communication. Charles R. Goldman, Professor of Limnology at the University of California Davis, initiated a long term study of Lake Tahoe in the 1960s based on a broad spectrum of measurements including nutrients, oxygen, plankton biomass, and photosynthesis. He noted that of all these limnological records, it was the reduction in Secchi depth through time that provided the most easily understood and convincing evidence for policy makers that stringent controls on watershed management were needed to protect the renowned clarity and blueness of the lake. The Secchi disk continues to be used routinely in lake studies (generally in combination with submersible optical instruments), and given its low cost and simplicity, it is also used in many parts of the world as part of public outreach activities and citizen monitoring programmes. The North American Lake Management Society (NALMS), for example, runs 'The Secchi Dip-In' every year that involves hundreds of lake shore residents and visitors throughout the USA and Canada.

Water colours

Forel took a special interest in how the colour of water differed among lakes and even among various parts of the same lake. He developed a liquid colour scale to classify different waters (today available as a smart phone app), and he undertook experiments to try to understand the reasons for these differences. Little could he imagine that water colour is now used in so many powerful ways to track changes in water quality and other properties of lakes, rivers, estuaries, and the ocean. A broad suite of optical technologies is increasingly available, ranging from profilers that can be lowered down into the lake to measure different spectral bands (including UV-radiation), to moored systems that automatically take measurements throughout the year, and satellites that continuously monitor changes in lake water colour from space.

Lakes have different colours, hues, and brightness levels as a result of the materials that are dissolved and suspended within them. The purest of lakes are deep blue because the water molecules themselves absorb light in the green and, to a greater extent, red end of the spectrum; they scatter the remaining blue photons in all directions, mostly downwards but also back towards our eyes. Some lake waters such as Lake Vanda in Antarctica and Crater Lake in Oregon are so deeply inky-blue that it almost seems as if dipping your hand in them would stain it indigo.

Algae in the water typically cause it to be green and turbid because their suspended cells and colonies contain chlorophyll and other light-capturing molecules that absorb strongly in the blue and red wavebands, but not green. However there are some notable exceptions. Noxious algal blooms dominated by cyanobacteria are blue-green (cyan) in colour caused by their blue-coloured protein phycocyanin, in addition to chlorophyll. Some village ponds in England have been called 'traffic light ponds' because they change from green to red over the course of the day. This is because their algae (classified in a group of motile species called Euglenophyta) contain red-coloured particles in addition to their photosynthetic pigments. In dim light and darkness, these red particles are hidden within the cellular interior, and the bright green chloroplasts are fully exposed towards the outside environment. When the sun is shining brightly, however, the red particles migrate outwards to mask the colour of the chloroplasts and protect the cells against solar damage. Some aquaculture fish pond managers have been startled to see their euglenophyte-containing ponds suddenly turn bright red in the sun. Red-coloured water may also be produced by other freshwater algae, for example *Uroglena* that forms red tides, cyanobacteria of the genus *Planktothrix* containing the red protein phycoerythrin, and the blood-red coloured alga *Haematococcus* that is common in garden bird baths throughout the world.

The yellowness that Forel observed in the inflows to Lake Geneva and its nearshore waters is due to the dissolved organic materials derived from the catchment. Specifically it is caused by the complex mixture of high molecular weight chemicals called 'humic acids': tea-like substances that are derived from the breakdown of leaf litter in the soils and that are then washed into the lake. These materials used to be called 'gelbstoff', which literally translates from the German into English as the imprecise term 'yellow stuff'. Forel was dead right when he said in 1895: 'What is the nature of these organic materials in the lake water? This question has not been sufficiently studied'. More than a hundred years later, this is still an active area of study in lake and ocean science. These days, the golden material is called 'coloured dissolved organic matter' (CDOM, pronounced 'see-dom'). This modern term again has a certain vagueness about it, concealing our still limited knowledge about the exact chemistry of this complex mixture.

One of the many interesting features of CDOM is that it strongly absorbs blue light, and absorbs even more strongly in the ultraviolet radiation (UV) part of the sunlight spectrum. It is therefore the natural sunscreen of lakes and rivers, protecting aquatic biota from harmful UV burn. The effect of CDOM on lake colour is all a matter of concentration. At highest concentrations, it absorbs sunlight at all wavelengths and the water is stained black, like espresso coffee. At lower, more typical, concentrations, the CDOM absorbs blue and blue-green light, leaving the water a golden brown colour, while at the lowest concentrations, CDOM absorbs the blue wavelengths, the water molecules themselves absorb the yellow to red wavelengths, and all that is left is the green waveband of light that is then perceived by our eyes. Forel proved this to himself by filtering brown, CDOM-rich swamp water, diluting it in clear lake water from Lake Geneva, and then filling a long, glass-bottomed tube that he could look through to the sky above; the water was transparent lime-green in colour, just as he had observed in the inshore region of Lake Geneva where inflowing streams mix into the lake.

Mysterious water

Water is such a common part our daily lives that we take it for granted, and certainly do not think of it as a chemical each time we turn on the tap or take a sip of something liquid. Yet it is a chemical substance with strange properties, some of which still elude a full explanation. These properties have enormous consequences for the physical and chemical nature of lakes, and for the aquatic life that lives within them.

At the heart of this strangeness is the water (H_2O) molecule itself and its tendency to aggregate in ever-changing clusters of different sizes and complexity. An electron is shared between each hydrogen atom and the oxygen atom, thereby forming a covalent bond that keeps the molecule together. But with eight times the positively charged protons than hydrogen, which has only one proton, the oxygen atom is the big brother in this relationship, and draws the shared, negative electron cloud slightly away from the hydrogen. The result is that the oxygen has a slight negative charge, while the two hydrogen atoms are left with slight positive charges. Opposites attract, so the molecules of water stick together, with the electrostatic O^-—H^+ interactions among them called hydrogen bonds (Figure 10). Each water molecule can hydrogen bond to a maximum of four others, and although the debate is ongoing, most H_2O molecules in liquid water appear to be dynamically linked together in pyramid-like clusters (tetrahedrons), with one of the oxygen atoms at the centre of the pyramid.

Another weird attribute of H_2O is its peculiar density versus temperature relationship. Solids tend to be more compact and dense relative to their liquid form, yet ice is exactly the opposite and floats on liquid water. This is because all of the H_2O

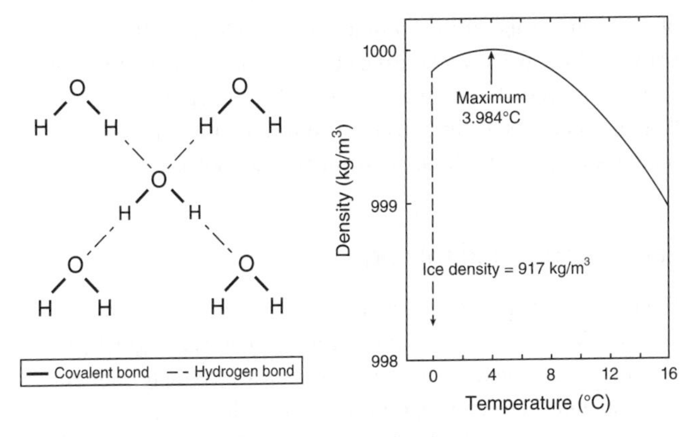

10. Hydrogen bonding and the strange density–temperature relationship for water.

molecules in ice are hydrogen bonded to four others with maximal distance among molecules in this crystal array. Once the ice melts, the liquid H_2O loses this expanded structure and full hydrogen bonding of all molecules; the molecules become more closely packed, resulting in greater density. This relaxing of structure continues with increased warming until around 4°C (3.984°C to be precise, at atmospheric pressure), which is the temperature of maximum density for pure water. With further warming above that, the increasing kinetic energy and motion of the water molecules result in an increasing average distance between them (although never to the extent of that in ice), and the density decreases (Figure 10).

Why is this density–temperature relationship so important? For Canadians and other northerners, it means that we can travel by ski, snow-shoe, and snowmobile over lakes in winter, confident in the knowledge that an impressively solid form of H_2O floats at the surface and will (hopefully) support us above the frigid liquid beneath. Of broader significance throughout the world, it also means that warm water floats on cold water, so that when the lake heats up in summer, a layer of warm water will float over the top of cold dense water at the bottom. The sharp temperature change between these two layers is called the 'thermocline', and the upper and lower layers that it separates are called the 'epilimnion' and 'hypolimnion', respectively. These two strata differ not only in temperature, but also in their chemistry and biology.

Lake seasons and mixing

From ice-covered waters to warm water floating on cold, the extent of layering of lakes, called 'stratification', varies greatly with season. At any one time, the layers may have strikingly different physical and chemical properties. For example, at Lake St-Charles, our drinking water reservoir in Quebec City, Canada, the thermocline in late summer lies at 7–10m depth (Figure 11). There is not only a sharp drop in temperature over this depth range, but also an abrupt decline in oxygen. The epilimnion is well charged in oxygen by exchange with the overlying atmosphere, but the thermocline acts as a barrier to that exchange, and the

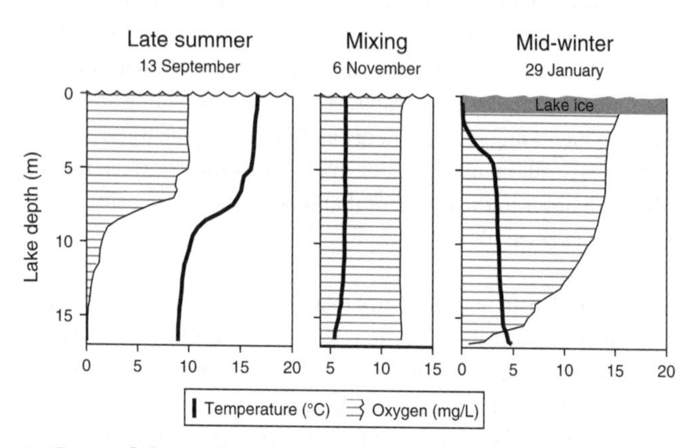

11. Seasonal changes in Lake St-Charles, the water reservoir for Quebec City. The bottom scale is for both temperature (thick lines, values in degrees Celsius (°C)) and oxygen (shaded areas, values in milligrams per litre (mg/L)).

bottom waters are completely depleted in this life-supporting gas. The hypolimnion of Lake St-Charles in late summer to early autumn is therefore a singularly unattractive place for fish such as trout, which only thrive in highly oxygenated waters.

With autumnal cooling of the epilimnion, the temperature difference between the top and bottom of the lake becomes much less, and there is no longer the strong density barrier against mixing between the surface and bottom layers. Additionally and most importantly, as the lake surface cools, the water becomes colder, denser, and sinks, giving rise to convective circulation that acts in concert with the wind-induced mixing. Eventually the entire water column mixes, called lake overturn, and the bottom waters are recharged with oxygen from the atmosphere (Figure 11). The solubility of all gases, including oxygen, is greater in cold water, and so by the end of this mixing period, the O_2 concentration increases to well above that in the warm summer epilimnion.

Temperate lakes in milder climates may be too warm to freeze in winter. These lakes are called 'monomictic' because they have only one, albeit long, period of vertical mixing, and atmospheric oxygen continues to recharge the lake as the water cools and mixes from autumn through winter. Many lakes of the world are monomictic, including lakes Biwa, Geneva, Tahoe, Titicaca, Taupo, and Maggiore, and lakes of the English Lake District.

The solubility of oxygen in water, even cold water, is not that great relative to the demand for it by chemical and biological processes, and in all lakes there is a precarious balance between income and expenditure columns of the ledger for this vital gas. This is nowhere more apparent than in northern, temperate lakes that have an ice cap in winter, such as Lake St-Charles (Figure 11). Although such lakes are well charged in oxygen due to pre-winter cooling and mixing, the winter ice and its associated snow cover cuts off light for oxygen generation by photosynthesis and eliminates the possibility of oxygenation from the overlying atmosphere. Decomposition meanwhile continues to consume oxygen, especially in the sediments, and can eventually drive the bottom waters to complete absence of oxygen (i.e. anoxia). Throughout winter, the lake is highly layered, this time with the colder (but less dense) water floating over the somewhat warmer water beneath. This is referred to as 'inverse stratification' because of the inverted temperature profile (cold over warmer), but the water density increases with depth, as it must to be in gravitational equilibrium.

In spring, the ice melts, Canadians pack away their snow boots, and the frigid surface layer of the lakes warms to the temperature of the waters beneath. With little difference in temperature and therefore density between the surface and deeper waters, the wind is able to mix and re-oxygenate the entire lake from top to bottom. Such lakes therefore have a second season of mixing, and are referred to as 'dimictic lakes', with two full water column mixing periods each year: autumn and spring. Unlike in autumn, however, spring mixing is rapidly dampened by the ongoing seasonal changes in temperature; once the surface water warms above 4°C it becomes less dense than the cold spring condition, and begins to float at the surface as an ever warming layer that impedes further mixing. For this reason, the spring mixing period in dimictic lakes can be brief and sometimes barely apparent relative to the prolonged period of mixing in autumn.

Lively waves at the surface

When the wind gently blows across a lake, it roughens the water surface to produce ripples. These wavelets are dragged upwards by the friction of the wind, and they subside into troughs as a result of the hydrogen bonding that pulls water molecules back down into the lake. They are referred to as 'capillary waves', to formally acknowledge that their restoring force is the capillary molecular interaction (i.e. surface tension) of the water. They have a maximum wavelength of 1.73cm and a period of less than a second. As the wind builds, the waves are dragged higher and now the restoring force is dominated by gravity. At winds above 25–30km per hour, or as the waves move into shallower depths inshore, the tops of the waves start travelling faster, overextend the base of the waves and break to produce whitecaps, causing intense mixing and oxygenation of the surface waters. Gravity waves have been observed up to 8m amplitude (i.e. trough to crest) in the North American Great Lakes, but lakes do not have the vast wind fetch of the ocean and most lake waves are less than 50cm in height.

At first impression, it seems that surface gravity waves have a large amount of energy that should mix the lake. The waves do give rise to movement below the surface, specifically a series of circular motions that decrease in diameter exponentially with depth, but they are not enough to fully mix the water column. This wave action causes the re-suspension of fine sediments in the littoral (i.e. inshore) region of the lake, which means that fine sediments only accumulate offshore, in deeper water, where the wave effects cannot penetrate. The depth threshold between re-suspension and sediment deposition is called the 'mud depositional boundary depth', and it depends on the wave height generated by storm events and on the slope of the inshore zone. Greater mixing of the lake, however, depends not on these 'lively waves', but rather on slower waves that are much less obvious to most lake visitors.

Slow waves at and below the surface

Forel noted in his autobiography that one of his favourite subjects of research was studying the slow, pendulum-like rocking motion of lakes. This phenomenon is so well known at Lake Geneva that the residents gave a name to it in the local Swiss-French dialect; for many centuries they have called it a 'seiche' (pronounced 'say-sh'), a term that is now used by scientists around the world to describe this ubiquitous feature of lakes. The most obvious aspect of seiches is the change in lake level, which rises and falls over a period of minutes to hours, particularly at the shore.

Forel made detailed observations of seiches at Lake Geneva and elsewhere with various types of continuous lake level recorders that he had conceived and installed, including a portable version. He was frustrated with his initial attempts to derive a mathematical theory of seiches, and noted with regret in his autobiography that while at college he had abandoned a useful class in differential calculus, as the teacher was particularly uninspiring (this itself is a lesson to all professors).

Forel did, however, have many other talents, including the enthusiasm and ability to network with other scientists throughout Europe and the world. He made contact with one of the leading physicists of the day, William Thomson, later named Lord Kelvin, who helped him simplify an initial unwieldy equation into the elegant form:

$$P = 2L / \sqrt{(gh)}$$

where P is the period of the rise and fall in lake level, L is the length of the lake, g is the gravitational constant, and h is the average depth of the lake. Forel correctly surmised that the seiche is a standing wave that extends across the full length of the lake, and that it could also be accompanied by lower amplitude, secondary waves.

So what is the origin of these oscillations? Forel rightly concluded that seiches begin by a consistent wind that blows and pushes the water towards one end of the lake (Figure 12). This 'set-up' condition results in the lake level being higher downwind and lower upwind, but it is not stable, and as soon as the wind stops, the water rocks back and overshoots in the opposite direction. This see-saw motion will continue until all the potential energy

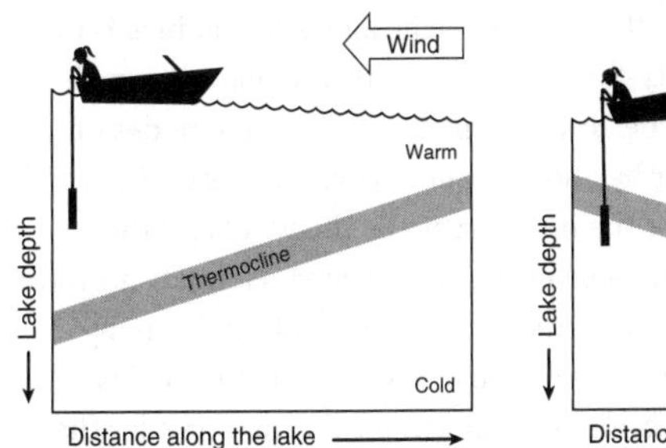

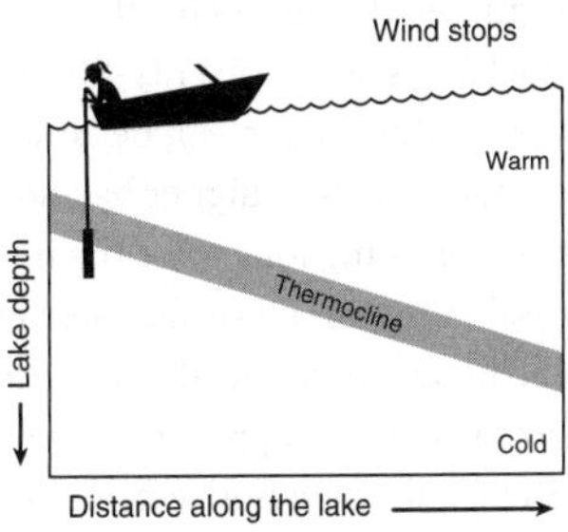

12. The surface seiche is caused by the wind pushing water to one end of the lake, and the resultant internal seiche can be detected as an oscillation of the thermocline.

associated with the initial displacement of the lake surface is finally dissipated, just as a pendulum finally comes to rest.

Forel would have been amazed to learn that the greatest significance of seiches lies well beneath the lake surface, at the depth of the thermocline where waves and mixing affect the transfer of oxygen and nutrients. He had evidence from his own work that the thermocline could change in depth over short periods of time, but he did not at the time make the link with the surface variations in water level. It was not until the classic work by Ernest M. Wedderburn and others in the lochs of Scotland that the nature of this 'internal wave' or 'internal seiche' came to be revealed. When the water piles up at the downwind end of a lake, the greater overlying mass of water in the epilimnion pushes the thermocline downwards. When the wind stress is relaxed, the thermocline rises up again, overshoots, and continues to oscillate until the set-up energy is dissipated.

Figure 12 captures the general idea of this wind-induced motion of the thermocline, but it needs some qualification. The vertical scale has been exaggerated and it does not express an important feature: that the internal seiche is slower and has a period that is much longer than that of the surface seiche. In Lake Geneva for example, the period for the surface seiche along the main axis of the lake is about seventy-four minutes, while the main internal wave (which again can be accompanied by additional modes of smaller waves of higher harmonics) has a period of three days, and it keeps going long after the initial setup by the wind and the dissipation of the surface seiche. The freshwater scientist shown in the figure would need to remain anchored on station for many hours to days to properly observe this slow moving internal wave. She would then notice from her underwater thermal profiles that the temperature at any specific depth gradually oscillates up and down, with the greatest variations in and near the thermocline.

Internal waves continue to be of intense interest to freshwater researchers, for many reasons. First, there is the question of size: they can be huge. Surface seiches are displacements of the air–water interface, and given the large difference in density of these two fluids, it takes a lot of wind energy to cause even a small increase in water level (thus potential energy) at the downwind end; for this reason, the amplitude of surface seiches is generally small, of the order of cm to tens of cm, although exceptions are known; for example, 5m seiches associated with storm surges on Lake Erie. For the internal wave, on the other hand, the density differences between the epilimnion and hypolimnion are small, and the same potential energy associated with a given surface displacement translates into a massive displacement at the level of the thermocline. During the set-up phase of internal waves in deep lakes such as Lake Tahoe, the bottom water can be displaced vertically by as much as 100m, as an 'upwelling' event near the upwind shore; this brings nutrients up into the photic zone, where they can stimulate algal production.

The motion of water in lakes and seas is affected by the rotation of the Earth, and internal waves are no exception: as the waves oscillate on the thermocline, they are deflected to the left in lakes of the Southern Hemisphere and to the right in lakes of the North Hemisphere. This so-called 'Coriolis Effect' is relatively weak, but in medium to large lakes it can be observed to modify the internal waves in two ways. First, it causes waves to be trapped and guided around the edge of the lake, in anticlockwise (Northern Hemisphere) or clockwise (Southern Hemisphere) direction. Lord Kelvin was the first to discover this phenomenon that is found in the atmosphere and ocean as well as in large lakes, and to define it formally. It is also fitting that the term 'Kelvin wave' makes reference to the physicist who helped Forel develop his theory of seiches (and perhaps Lord Kelvin was stimulated in his own thinking about waves during these discussions), for in many lakes this is a feature with major consequences. In Lake Biwa, Japan, for example, the Kelvin wave can be so highly energized by the strong winds of the typhoon season that it sometimes attains the surface, bringing cold, nutrient-rich waters into the upper layers as it circulates around the lake.

The second type of internal wave affected by the Coriolis force occurs offshore, in the main body of the lake, and is called a 'Poincaré wave', named after the brilliant French mathematician and theoretical physicist Henri Poincaré. These waves can similarly be of high amplitude. The example shown in Figure 13 for Lake Ontario, Canada–USA, is from temperature measurements at a fixed sampling station over a five-day period. The wave-like nature of the observations is striking, as is the amplitude of oscillation,

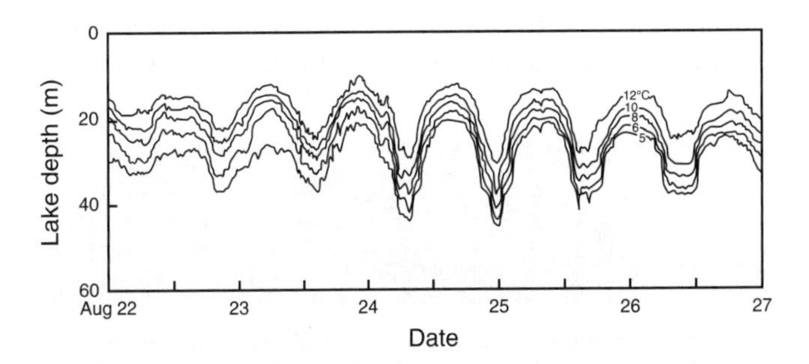

13. Poincaré waves on the thermocline of Lake Ontario.

which reached 25m from trough to crest. Poincaré waves have a much shorter period than the inshore Kelvin waves, but still much longer than the surface seiche; in Lake Ontario this period is sixteen hours (Figure 13) versus ten days for the Kelvin wave and five hours for the surface seiche along the main axis of the lake.

Lake biologists and biogeochemists all have a special interest in internal waves because these motions play a decisive role in mixing the water between layers, which, as seen in Figure 11, can differ greatly in temperature and oxygen, as well as in many other properties. The wave-induced, horizontal movement of water extends to the lake floor: as the water mass slides backwards and forwards within the lake basin, an oscillating turbulent current moves across the sediments and can bring particles up into suspension into a bottom zone called the 'benthic boundary layer'. This water flow and mixing drives oxygen into the sediments, and it accelerates exchanges of other chemicals such as nitrogen and phosphorus between the lake bottom and the overlying water. Of special relevance to the plankton, the tilting of the thermocline during set up and subsequent oscillations brings deep water closer to the surface, and it exposes that water to surface mixing. This entrains nutrients from depth into the photic zone, and gives rise to horizontal flows that help mix and homogenize the lake.

And then there are 'billows' (Figure 14). The internal seiche generates water movements that are in opposing directions above and below the thermocline. This creates friction and a pressure

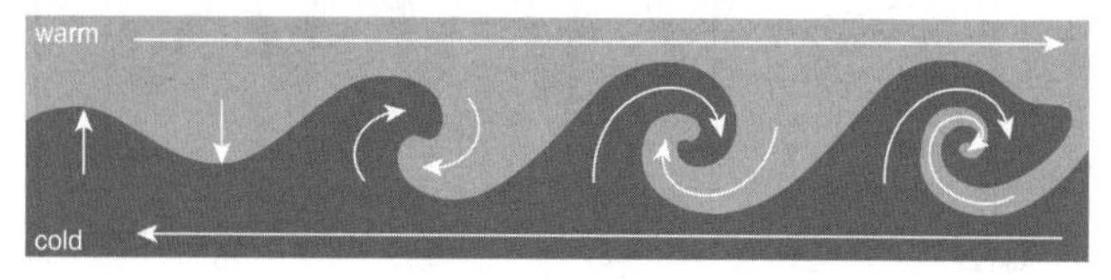

14. Billows across the thermocline. These begin as progressive waves that roll up, break, and mix materials between the epilimnion and hypolimnion.

gradient that results in short-period, progressive waves that propagate along the thermocline, superimposed upon the internal, standing wave. These higher frequency waves typically have periods of about a hundred seconds, wavelengths of 10–50m, and amplitudes of 0.05–2m. Most importantly, the waves can roll up and break, just like waves at the surface, opening a temporary window in the thermocline barrier that allows the exchange of heat, nutrients, oxygen, and particles. These billowing effects in a layered fluid are called 'Kelvin–Helmholtz instabilities' (Lord Kelvin yet again, this time in the company of the famous German physicist, Hermann von Helmholtz), and they can sometimes be observed in the sky as swirls of cloud that form as warm air is mixed into cold. The onset of these billows depends on the velocity gradient across thermocline. They can occur near the lake margin, giving rise to increased nutrient supply and primary productivity there, or this mixing can occur mid-lake, causing lake-wide algal growth.

Currents in the lake

The oscillating flows associated with surface and internal seiches are only a subset of the dazzling variety of water mass movements that occur within lakes. Forel pointed out that at the largest dimension, at the scale of the entire lake, there has to be a net flow from the inflowing rivers to the outflow, and he suggested that from this landscape perspective, lakes might be thought of as enlarged rivers. Of course, this riverine flow is constantly disrupted by wind-induced movements of the water. When the wind blows across the surface, it drags the surface water with it to generate a downwind flow, and this has to be balanced by a return movement of water at depth. This explains the observations by the Lake Geneva fisherman that their deep nets could be displaced in a direction that was opposite to the prevailing wind.

In large lakes, the rotation of the Earth has plenty of time to exert its weak effect as the water moves from one side of the lake to the other. As a result, the surface water no longer flows in a straight line, but rather is directed into two or more circular patterns or gyres that can move nearshore water masses rapidly into the centre of the lake and vice-versa. Gyres can therefore be of great consequence, short-circuiting water and its associated content, including pollutants and even toxic algae, from one location to another. They give rise to a striking pattern of water flow; for example, at Lake Biwa, Japan, which is said to have the most beautiful gyres in the world (Figure 15). Unrelated to the Coriolis Effect, the interaction between wind-induced currents and the shoreline can also cause water to flow in circular, individual gyres, even in smaller lakes.

At a much smaller scale, the blowing of wind across a lake can give rise to downwind spiral motions in the water, called 'Langmuir cells'. These were first observed by the distinguished American

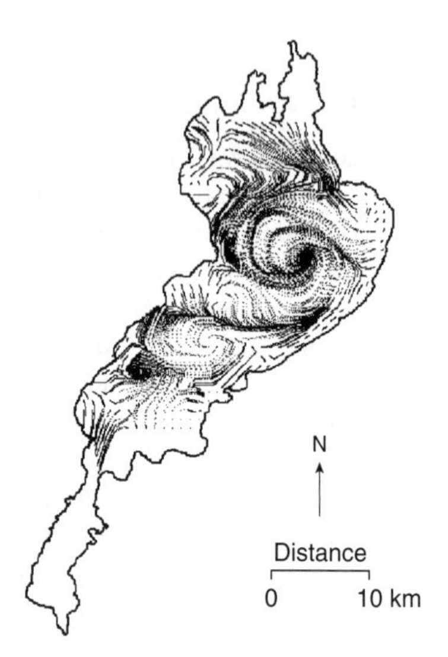

15. The gyres of Lake Biwa, Japan, based on measurements with an Acoustic Doppler Current Profiler (ADCP) on board the research vessel Hakkengo.

scientist Irving Langmuir in the Sargasso Sea, where adjacent, counter-rotating spirals of water movement concentrate floating material at the surface. In that region of the ocean, the material is seaweed of the genus *Sargassum*, which forms long parallel lines of accumulation. These circulation features are commonly observed in lakes, where the spirals progressing in the general direction of the wind concentrate foam (on days of white-cap waves) or glossy, oily materials (on less windy days) into regularly spaced lines that are parallel to the direction of the wind.

Density currents must also be included in this brief discussion of water movement, for they play a major role in many if not most of the world's lakes, including Lake Geneva. Just like stratification, this feature is a result of the density–temperature relationship of water. Cold river water entering a warm lake will be denser than its surroundings and therefore sinks to the bottom, where it may continue to flow for considerable distances. In Lake Geneva, the cold, sediment-laden water of the upper Rhône River enters the lake, immediately plunges and moves many kilometres along the lake floor where it has cut a deep ravine called the Rhône Canyon. Measurements in that canyon using multi-beam echo-sounders indicate that in some places the sediments can be eroded by several metres each year by this process, while in other locations the transported sediment builds up as bottom deposits. In this way, the submarine canyon is an ever-changing, sinuous valley at the bottom of the lake, and a conduit and etching track for its glacial inflow. Density currents contribute greatly to inshore–offshore exchanges of water, with potential effects on primary productivity, deep-water oxygenation, and the dispersion of pollutants.

Chapter 4
Life support systems

> No serious analysis, to my knowledge, has indicated any lake water completely free of microbes...
>
> F. A. Forel

François Forel was on the right track when he acknowledged the widespread presence of microbes, but little could he suspect the astonishing variety and abundance of microscopic life that underpins the ecology of natural waters. A cup of lake water scooped from even the clearest of lakes will contain a populous yet invisible living world: perhaps 100,000 photosynthetic cells, ten million bacterial cells, and a hundred million 'wild viruses' in suspension, all unapparent to the naked eye. Freshwater ecologists have long puzzled over the co-existence of so many different types of algal cells in the homogenous surface waters of lakes. G. Evelyn Hutchinson, one of the most renowned of lake scientists, called this the 'paradox of the plankton': why is it that one species does not simply drive all others to extinction by outcompeting them for the limited resources? With the advent of new molecular and biochemical techniques, this spectacular diversity of microbes has become even more apparent. Just as we now realize that the ensemble of microbes that live on and within our bodies, the 'human microbiome', can greatly affect our state of health, the 'aquatic microbiome' is central to the healthy functioning of lake ecosystems and their responses to environmental change.

Solar-based economies

Almost all ecosystems on Earth depend on the input of energy from the sun, either directly for photosynthesis in the present, or indirectly through the past accumulation of photosynthetic biomass, and its subsequent use by microbes and animals. The recycling of old plant material is especially important in lakes, and one way to appreciate its significance is to measure the concentration of CO_2, an end product of decomposition, in the surface waters. This value is often above, sometimes well above, the value to be expected from equilibration of this gas with the overlying air, meaning that many lakes are net producers of CO_2 and that they emit this greenhouse gas to the atmosphere. How can that be?

To find an answer to this question we need to move outside the boundary of the water-filled basin. Lakes are not sealed

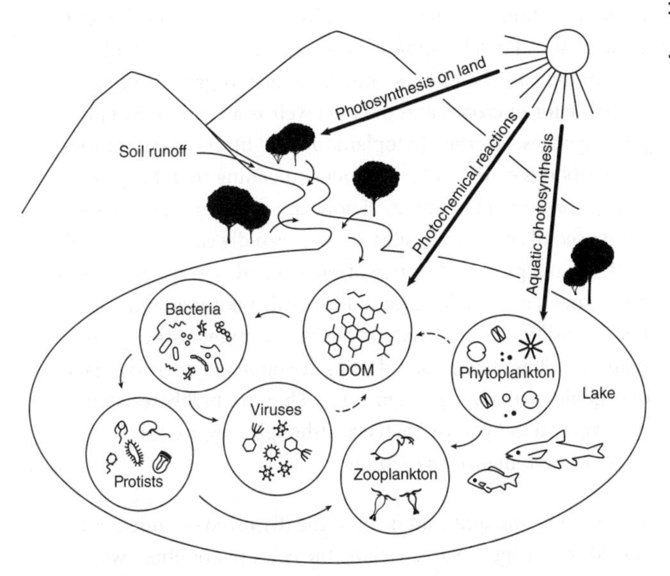

16. From sunlight to diverse microbes and the aquatic food web.

microcosms that function as stand-alone entities; on the contrary, they are embedded in a landscape and are intimately coupled to their terrestrial surroundings. Organic materials are produced within the lake by the phytoplankton, photosynthetic cells that are suspended in the water and that fix CO_2, release oxygen (O_2), and produce biomass at the base of the aquatic food web. Photosynthesis also takes place by attached algae (the periphyton) and submerged water plants (aquatic macrophytes) that occur at the edge of the lake where enough sunlight reaches the bottom to allow their growth. But additionally, lakes are the downstream recipients of terrestrial runoff from their catchments (Figure 16). These continuous inputs include not only water, but also subsidies of plant and soil organic carbon that are washed into the lake via streams, rivers, groundwater, and overland flows.

The organic carbon entering lakes from the catchment is referred to as 'allochthonous', meaning coming from the outside, and it tends to be relatively old because it was produced by plants on land in the past. In contrast, much younger organic carbon is available for microbes and the food web as a result of recent photosynthesis by the phytoplankton and littoral communities; this carbon is called 'autochthonous', meaning that it is produced within the lake. This young, dissolved, organic matter is mostly composed of small molecules and it is rapidly consumed by lake bacteria as a preferred source of carbon and energy. On the other hand, much of the allochthonous organic matter is composed of humic and fulvic acids that are derived from terrestrial plant material. These tea-coloured acids are large polymers composed of organic carbon rings (Figure 16). They are highly resistant to decomposition by most bacteria, although they are subject to decay by certain groups of fungi.

It used to be thought that most of the dissolved organic matter (DOM) entering lakes, especially the coloured fraction, was unreactive and that it would transit the lake to ultimately leave unchanged at the outflow. However, many experiments and field observations have shown that this coloured material can be partially broken down by sunlight. These photochemical reactions result in the production of CO_2, and also the degradation of some of the organic polymers into smaller organic molecules; these in turn are used by bacteria and decomposed to CO_2. This sunlight-driven chemistry begins in the rivers, and continues in the surface waters of the lake. Additional chemical and microbial reactions in the soil also break down organic materials and release CO_2 into the runoff and ground waters, further contributing to the high concentrations in lake water and its emission to the atmosphere. In algal-rich 'eutrophic' lakes there may be sufficient photosynthesis to cause the drawdown of CO_2 to concentrations below equilibrium with the air, resulting in the reverse flux of this gas, from the atmosphere into the surface waters.

Precarious oxygen

There is a precarious balance in lakes between oxygen gains and losses, despite the seemingly limitless quantities in the overlying atmosphere. This balance can sometimes tip to deficits that send a lake into oxygen bankruptcy, with the O_2 mostly or even completely consumed. Waters that have O_2 concentrations below 2mg/L are referred to as 'hypoxic', and will be avoided by most fish species, while waters in which there is a complete absence of oxygen are called 'anoxic' and are mostly the domain for specialized, hardy microbes. In the warm waters of the Amazon, for example, the oxygen balance can shift to anoxia during the night, and the fish must therefore migrate into shallower oxygenated waters or out to the river—where predatory species such as large catfish are lying in wait. In many temperate lakes, mixing in spring and again in autumn are the critical periods of re-oxygenation from the overlying atmosphere. In summer, however, the thermocline greatly slows down that oxygen transfer from air to deep water, and in cooler climates, winter ice-cover acts as another barrier to oxygenation. In both of these seasons, the oxygen absorbed into the water during earlier periods of mixing may be rapidly consumed, leading to anoxic conditions.

Part of the reason that lakes are continuously on the brink of anoxia is that only limited quantities of oxygen can be stored in water because of its low solubility. The concentration of oxygen in the air is 209 millilitres per litre (ml/L; 20.9 per cent by volume), but cold water in equilibrium with the atmosphere contains only 9ml/L (0.9 per cent). This scarcity of oxygen worsens with increasing temperature (from 4°C to 30°C the solubility of oxygen falls by 43 per cent), and it is compounded by faster rates of bacterial decomposition in warmer waters and thus a higher respiratory demand for oxygen. Additionally, the microbial demand for this sparse resource can be excessive because lakes are intense sites of decomposition for the entire landscape, with large populations of oxygen-consuming bacteria that are fuelled by organic carbon from the watershed as well as within-lake sources (Figure 16).

Identifying the invisible

In his description of the microbiology of Lake Geneva, Forel in 1904 pointed out that although bacteria were abundant everywhere, they were not to be feared: 'the immense majority of these tiny life forms are entirely innocent'. Innocent perhaps, but certainly not inconsequential given their massive population size as well as diversity of species and job descriptions. Up until recently, identifying the bacteria present and their specific roles in lake ecosystems was simply impossible because most species could not be brought into culture, and most cannot be identified or differentiated by just looking at them under a microscope. These days, however, the application of methods based on nucleic acid sequencing, specifically DNA to examine the genes present in the community and RNA to determine what genes are being read or 'expressed' to produce proteins, is allowing lake samples to be analysed without going through the problematical culture step.

These analyses are revealing an extraordinarily diverse community of active microbiota and microbial functions that collectively make up the lake microbiome. Many groups and species are still poorly understood, and new microbes and processes continue to be discovered every year.

Microbiomes are composed of four microbial components, and there is immense species and functional diversity within each of them. The smallest, most abundant, and probably most diverse are viruses. These miniaturized parasites attack and reprogramme cells to produce more viral particles, and they typically range in size from 20 to 200 nanometres (nm—one millionth of a millimetre). Each cellular microbe has its own set of pathogenic viruses, and because bacteria are the most abundant cells in the microbiome, the majority of naturally occurring 'wild viruses' in the water are bacterial parasites, called 'bacteriophages' or simply 'phages'. Others, however, attack various other components of the microbial food web, which may affect seasonal variations in this food web to an extent that is presently not well understood. A group called 'giant viruses' (Mimivirus and relatives) has attracted great interest because of their size (greater than 250nm) and their role as parasites of amoebae and algae. There are also viruses that attack aquatic animals, such as the fish parasite 'infectious haematopoietic necrosis virus' (IHNV). This infects trout and salmon species and can cause high rates of mortality in fish farms.

Once the progeny of viruses are fully assembled during the reproduction phase, the host cell is then induced to 'lyse' (i.e. burst open) and release them, in the process disgorging other cellular materials into the water. These materials are choice substrates for uptake and growth by bacteria that have so far eluded attack, but this success may be short-lived before these other bacteria are themselves infected by viruses, or are eaten by protists, who in turn are eaten by zooplankton (Figure 16). This shunting of organic carbon from bacteria via viral lysis to other bacteria then to protists is referred to as the 'viral shunt', and in some lake and ocean environments it may account for more than 10 per cent of the total carbon flow. Viruses play an additional role in transferring chunks of DNA among hosts, which may confer new or modified gene functions that can then be passed onto future generations, although only if the host is lucky enough to avoid viral lysis in this continuous microbial warfare.

The second component of the microbiome is the bacteria themselves, and many branches (phyla) of the bacterial 'tree of life' are well represented in lakes. One way to appreciate their abundance and diversity is to dye the cells in a sample of lake water with a fluorescent stain, filter the cells onto a membrane, and look at the membrane under a fluorescence microscope. The sample will light up like the Milky Way, with thousands of star-bright, fluorescing cells that vary in size and form: mostly spheres (i.e. cocci) but also oblong (i.e. rods), spiral, kidney-shaped, and filamentous forms. It will be seen that most of the cells in this microbial constellation are extremely small, in the range 200–400nm in diameter, too small for Forel to have detected with his standard microscope. These so-called 'ultramicrobacteria' have the advantage of exposing a large cell surface, relative to their volume, to the lake environment. This maximizes their chances of encountering and absorbing (through specialized 'transport' proteins on their outer membranes) the organic molecules and nutrients that occur at dilute concentrations in the lake water.

The most common phylum of pelagic bacteria is the Proteobacteria, with three notable subphyla in lakes; alpha-, beta-, and gammaproteobacteria. Betaproteobacteria are the most abundant, constituting up to 70 per cent of the total number of cells in the plankton. They include the aptly named lake inhabitant, *Limnohabitans*, which appears to be able to grow rapidly on organic materials released by phytoplankton and thereby outpace the dual pressure of grazing and viral attack. Another common betaproteobacterium found throughout the world's lakes is

Polynucleobacter, which has the capacity to utilize complex organic materials, including breakdown products of humic acids flowing in from the catchment. A betaproteobacterium that plays an especially important role in the nitrogen cycle is the nitrifier *Nitrosomonas*, which oxidizes ammonium (NH_4^+) to nitrite and in the process consumes large quantities of oxygen.

Gammaproteobacteria is a bacterial group that mostly occurs in marine waters, but there are two subgroups that deserve special attention in lakes. The family Methylococcaceae includes several genera that use methane as their carbon and energy source; these include *Methanococcus* and *Methylobacter*, often found near anoxic environments where methane is being generated, such as on the surface of lake sediments. Another family is the Enterobacteriaceae, and its most infamous member is *Escherichia coli*. This species, usually abbreviated as *E. coli*, is named after the Austrian pediatrician Theodor Escherich who first isolated it from the faeces of an infant. Most *E. coli* are not pathogenic (although there are dangerous exceptions), but they are used in monitoring of drinking water and bathing areas on lakes as an indicator of human faecal contamination. Apart from infectious strains of *E. coli*, this contamination may include other pathogenic microbes causing water-borne diseases such as cholera, hepatitis, typhoid, and gastrointestinal illnesses.

Most of the bacterial species in lakes are decomposers that convert organic matter into mineral end products, in particular carbon dioxide, ammonium, phosphate, and hydrogen sulfide (H_2S). In addition to this essential composting and recycling role, some bacteria specialize in transforming inorganic molecules as a source of energy (such as the nitrifiers), and there are others that depend upon sunlight. The most notable of the latter are apparent under the fluorescent microscope as brightly fluorescing red-orange balls amongst the constellation of stained cells. These are 'picocyanobacteria' and although they are larger than the decomposers, they are still pretty small, around 2 micrometres (μm) or less. They would have escaped detection by Forel with a standard microscope, yet they are probably the most abundant photosynthetic cells in Lake Geneva, as in most lakes and oceans. Their red-orange glow in a fluorescence microscope is due to their blue and red protein pigments that, along with chlorophyll, absorb light and fluoresce.

Archaea (or archaeons), the third constituent of microbiomes, share certain features with bacteria in that their cells are small, non-descript, and 'prokaryotic'; that is, they lack a nucleus and other cellular structures that are typical of more advanced 'eukaryotic' cells, including our own. This simplicity, however, belies a set of unusual features that makes them genetically and biochemically distinct, and microbiologists classify archaeons apart from bacteria and eukaryotes as the 'third domain of life'. Some of these microbes play important biogeochemical roles in natural waters, such as the production of methane and the oxidation of ammonium.

The last but certainly not least component of the lake microbiome is the assemblage of 'microbial eukaryotes' with their more complex nucleated cells. Also called 'protists', these include two major groups that were historically separated by function: photosynthetic protists or algae, which capture sunlight for photosynthesis and use CO_2 as their carbon source; and colourless protists that are fuelled by organic molecules that they absorb from the lake water or extract from bacteria. Even the healthiest of algae leak some of their products of photosynthesis into the water during their growth and reproduction, and much more of this organic matter is released when the algal cells are broken up by zooplankton or burst by viruses. This organic carbon would be lost from the food web were it not for its uptake by bacteria and then capture of those bacteria by protists. This carbon recovery process is referred to as the 'microbial loop', and some of this carbon and energy can then move up the food web via zooplankton, ultimately to fish (Figure 16).

Not so very long ago, biologists had a clearly defined view of the living world that was based on carbon and energy sources: inorganic versus organic, photosynthesis versus feeding, plants versus animals. However, protists have little respect for this scientific clarity because many are able to alternate between plant and animal modes of life. Such species that depend upon this mixed combination of energy sources are referred to as 'mixotrophs', and they occur commonly in most lake waters. This strategy allows them to exploit the best of both worlds by harnessing sunlight with their photosynthetic pigments and by exploiting the local resources of pre-made organic compounds. Their capture of bacteria is especially effective, for these minute cells have already done the hard work of scavenging organic molecules and nutrients from the surrounding lake water. The bacterial cells thereby provide concentrated, high-energy food packages for mixotrophic protists, as well as for non-photosynthetic protists such as colourless flagellates and ciliates.

Cycles that matter

Lake microbiomes play multiple roles in food webs as producers, parasites, and consumers, and as steps into the animal food chain (Figure 16). These diverse communities of microbes additionally hold centre stage in the vital recycling of elements within the lake ecosystem, in particular by their activities of oxidation (loss of electrons) and reduction (gain of electrons). These biogeochemical processes are not simply of academic interest; they totally alter the nutritional value, mobility, and even toxicity of elements. For example, sulfate is the most oxidized and also most abundant form of sulfur in natural waters, and it is the ion taken up by phytoplankton and aquatic plants to meet their biochemical needs for this element. These photosynthetic organisms reduce the sulfate to organic sulfur compounds, and once they die and decompose, bacteria convert these compounds to the rotten-egg smelling gas, H_2S, which is toxic to most aquatic life. In anoxic waters and sediments, this effect is amplified by bacterial sulfate reducers that directly convert sulfate to H_2S. Fortunately another group of bacteria, sulfur oxidizers, can use H_2S as a chemical energy source, and in oxygenated waters they convert this reduced sulfur back to its benign, oxidized, sulfate form.

The carbon cycle is at the heart of ecosystem functioning (Figure 17). Microbes are responsible for many of the key transformations, but mineral chemistry also plays a major role. Inorganic carbon enters the lake in three different forms: gaseous CO_2, bicarbonate ions (HCO_3^-), and carbonate ions (CO_3^{2-}). Carbonate is mainly derived from the weathering of limestone (calcium carbonate) and dolomite (calcium and magnesium carbonate) in the watershed, while the CO_2 enters from the atmosphere, inflows, and from respiration, mostly associated with the bacterial decomposition of organic matter. These three forms are in chemical equilibrium with each other via the equation:

$$2H^+ + CO_3^{2-} \rightleftharpoons H^+ + HCO_3^- \rightleftharpoons H_2CO_3 \rightleftharpoons CO_2 + H_2O$$

This equation is central to the pH balance of the lake, because it means that acids entering the water will be rapidly neutralized: the carbonate and bicarbonate ions will take up acid protons (H^+). However, this acid neutralizing capacity (or 'alkalinity') varies greatly among lakes. Many lakes in Europe, North America, and Asia have been dangerously shifted towards a low pH because they lacked sufficient carbonate to buffer the continuous inputs of acid rain that resulted from industrial pollution of the atmosphere. The acid conditions have negative effects on aquatic animals, including by causing a shift in aluminium to its more soluble and toxic form Al^{3+}. Fortunately, these industrial emissions have been regulated and reduced in most of the developed world, although there are still legacy effects of acid rain that have resulted in a long-term depletion of carbonates and associated calcium in certain watersheds.

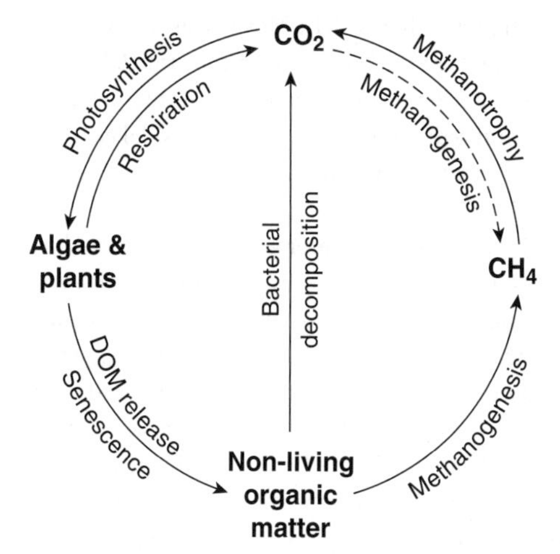

17. The aquatic carbon cycle.

CO_2 is continuously removed by photosynthetic microbes, the phytoplankton, as well as by water plants. To make up for that shortfall and maintain equilibrium, the inorganic carbon reactions are driven to the right to replace the consumed CO_2, and in the process they consume protons and thereby cause an increase in pH. At higher pH values and especially in warm water, the carbonate is no longer soluble and it precipitates out as a chalky coloured suspension. Astronauts on board the International Space Station have taken some striking pictures of such events, called 'whitings', in the North American Great Lakes.

Methane (CH_4) is the second gas of great interest in the aquatic carbon cycle (Figure 17), especially because it is a greenhouse gas like CO_2, but with more than twenty times the greenhouse warming potential per molecule. The generation of methane (i.e. methanogenesis) primarily takes place in anoxic environments by a variety of specialized archaea, the third domain of cellular life in the lake microbiome. Some of these microbes use CO_2 (the dashed line in Figure 17), while others use small organic molecules to produce the highly reduced CH_4. Mostly this takes place in the black-ooze sediments at the bottom of organic rich lakes, but it may also occur in bottom waters that have lost their oxygen, or in both the water and sediments within ice-covered lakes that become completely anoxic in winter.

Some impressive examples of winter gas production can be found in the thermokarst lakes that occur in great abundance across the Arctic tundra. These ecosystems are enriched by soil organic carbon that enters the lakes from thawing and eroding permafrost. The bacterial decomposition of that carbon causes rapid oxygen loss soon after freeze-up, and these ice-capped anoxic waters are ideal bioreactors for methane production by archaea. Making a hole through the ice will release methane bubbles that have accumulated over winter, and this vented gas can be set alight to produce a spectacular burst of flame above the snow and lake ice.

Organic molecules contain carbon in its reduced form, and the most reduced of all is methane. Oxidation back to CO_2 completes the cycle (Figure 17). In the case of most of the particulate and dissolved organic matter, derived mostly from dead algae and plants, this composting role is undertaken by many species of bacteria that can break down and derive energy from a great variety of organic substrates. For methane oxidation, this is the preserve of a more limited number of bacterial specialists called 'methanotrophs'. These bacteria are normally confined to a narrow interface zone where there is both methane and oxygen. A notable exception is in thermokarst lakes, where re-oxygenation from the air after the ice melts in summer offers a paradise of methane plus oxygen for methanotrophs. As might be expected, these microbes make up an unusually large fraction, sometimes more than 10 per cent, of all the bacteria in these permafrost thaw waters during summer.

Nitrogen cycling has a much greater complexity relative to the carbon cycle in terms of the different types of molecules and ions, their states of oxidation, and the array of microbial specialists that keep the cycle spinning (Figure 18). Nitrogen is the most abundant gas in the atmosphere, and also in lake water, but the triple bond of the N_2 molecule is extremely stable and difficult to break. Some cyanobacteria are capable of this enzymatic feat, notably certain planktonic bloom-formers such as *Dolichospermum* (previously called *Anabaena*) and bottom-dwelling forms such as *Nostoc*, which produces jelly-like sheets and balls. In general, however, N_2-fixation is not a major source of nitrogen for lake ecosystems. For example, in Lake Mendota, Wisconsin, one of the most studied lakes in the world, beginning with the classic work of pioneer limnologists Edward A. Birge and Chancey Juday, blooms of cyanobacteria occur each year, including nitrogen-fixing species. However, this biological fixation of atmospheric N_2 appears to account for less than 10 per cent of all nitrogen entering the lake.

Most of the nitrogen entering Lake Mendota, and lakes in general, is derived from their watersheds via inflows, and from their airsheds (the local or regional atmosphere) in rain, snow, and

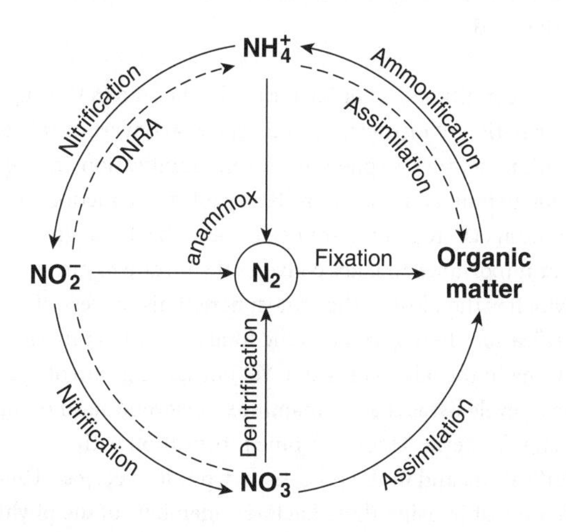

18. The aquatic nitrogen cycle.

wind-blown dust. This nitrogen arrives as nitrate, ammonium, and a variety dissolved and particulate organic forms. Some of this nitrogenous material is taken up by the phytoplankton during their cell growth (the assimilation processes in Figure 18), and when the cells die, some of this organic matter is eventually decomposed to organic nitrogen and ammonium ('ammonification').

Several processes lead the cycle back to nitrogen gas. Apart from being the preferred nitrogen for uptake by phytoplankton, ammonium (NH_4^+) can also be oxidized by a group of specialized bacteria (and some archaea) to nitrite (NO_2^-), and by other bacteria then to nitrate (NO_3^-). Some nitrifiers called 'complete ammonium oxidizers' (abbreviated comammox) can convert NH_4^+ all the way through to NO_3^-. As indicated by these chemical formulae, nitrifying bacteria have a gourmand appetite for oxygen: three atoms of oxygen are needed for each atom of nitrogen to be fully oxidized to nitrate. In the bottom waters of eutrophic lakes, the precarious oxygen balance can be tipped to complete anoxia in part because of this extremely large demand.

In anoxic conditions, other bacteria take the lead in the ongoing transformation of nitrogen. Certain species convert nitrate to ammonium, a process called nitrate ammonification or, in even more longwinded terms, dissimilatory nitrate reduction to ammonium (DNRA in Figure 18). Another bacterial group of great importance to lakes converts the nitrate to nitrogen gas, which is then lost to the atmosphere. This process of denitrification thereby reduces the total amount of nitrogen remaining in the lake ecosystem. Yet one more group of specialists on the N-cycle list is that of anammox (anaerobic ammonium oxidizing) bacteria. These combine nitrification with denitrification, and in the process they produce N_2 gas. There is much interest in using these bacteria (members of the phylum

Planctomycetes) for wastewater treatment because they can convert nitrogen-containing waste to N_2 that is vented to the air, and they have little requirement for organic carbon supplements, unlike denitrifying bacteria.

The biogeochemical phosphorus cycle comprises another set of oxidation and reduction processes that are immensely important for lake ecosystems. This element is often a limiting factor for algal growth, and is a major cause of lake enrichment by domestic wastes and fertilizers from agriculture. Unlike carbon and nitrogen, phosphorus has no gaseous form, and it arrives from catchment rocks and soils as suspended particulate matter and in dissolved forms such as orthophosphate and dissolved organic phosphorus. Phosphorus is best measured as 'total phosphorus' (TP), the sum of all dissolved and particulate forms, and TP concentrations range from less than 10 parts per billion (ppb) in clear oligotrophic lakes to 100ppb in algal-rich, eutrophic lakes.

Huge quantities of phosphorus are stored in lake sediments, but while most of this is locked away from the overlying water, some it can be mobilized and released under the right conditions. This was first demonstrated by the eminent limnologist Clifford H. Mortimer in a classic experiment using sediments from the bottom of Lake Windermere, in the English Lake District. He placed this mud on the bottom of an aquarium, overlaid the mud with lake water, and then measured the chemical changes associated with oxygen loss. Once the oxygen was depleted, there was an outpouring of dissolved iron and phosphate from the mud into the overlying water. This same effect can now be demonstrated with much greater resolution using micro-electrodes. In an experiment with sediments from Lake Erie, the shallowest waterbody of the North American Great Lakes, profiles with a novel phosphate electrode showed a rise in concentration in the surface sediments and in the overlying water by a factor of 10,000 after the shift from oxygenated to anoxic conditions (Figure 19).

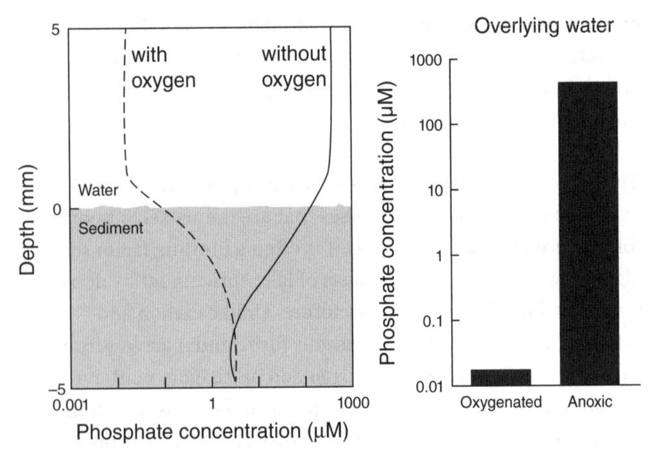

19. Phosphate release from Lake Erie sediments under oxygenated and anoxic (without oxygen) conditions.

Several mechanisms are responsible for these chemical changes at the sediment surface. Mortimer correctly surmised that much of the phosphorus in oxygenated sediments is attached (adsorbed) to insoluble iron, or ferric (Fe III) hydroxy-oxides. Under anoxic conditions these oxides are reduced to the ferrous (Fe II) form, which is soluble, and this allows the phosphate to be released into the overlying water. Furthermore, bacterial reduction of sulfate during anoxia produces sulfide that reacts and binds with the iron, resulting in less ferric hydroxy-oxide for phosphate adsorption. Additionally, oxygenated sediments may have a surface cap of bacteria that sequester phosphorus internally as polyphosphate granules, which may be released under anoxic conditions. There are other variables such as aluminium, calcium, organic carbon, and pH that strongly influence this complex biogeochemistry, and not surprisingly, there are large differences in the magnitude of this effect among lakes, including no response to anoxia in some. In many eutrophic lakes, however, once this oxic–anoxic threshold is crossed (Figure 19), phosphorus that was derived from the catchment over years to decades earlier is no longer locked in the sediments, and its release can accelerate enrichment and slow down attempts to improve lake water quality.

Zones of production

With the methods available to him at the time, it was impossible for Forel to be aware of the diversity of teeming microbes that support the natural cycles of lakes, and only now are we beginning to realize the genetic richness of these species and the complex network relationships among them. Forel was aware, however, that lakes can be divided into two zones that sharply differ in terms of their photosynthetic communities at the base of the aquatic food web. In the inshore (or 'littoral') zone, the primary producers are mostly aquatic plants, some rooted in the sediments and fully submerged, and others with leaves that float or extend out of the water. In shallow lakes and ponds, these water plants or 'aquatic macrophytes' and their associated microbes may dominate the overall biological production of the ecosystem. Forel was impressed by the luxuriant growth of these plant communities in Lake Geneva, noting in 1904 how 'they form true underwater forests, as picturesque, mysterious and attractive as the most beautiful forests of our mountains'. These provide important habitats for aquatic animals, as well as trapping nutrients and affecting water currents.

Offshore, in the limnetic or pelagic zone, beauty is more at the microscopic level, and the sweep of a plankton net though the water will yield a wonderfully diverse collection of pigmented cells that vary in size, shape, and colour. This algal plankton or phytoplankton captures light for photosynthesis, and their growth extends down to the bottom of the photic zone. This also demarcates the bottom and therefore offshore extent of the littoral zone, although individual plant species may be limited in their distribution by other factors such as water pressure, grazing animals such as crayfish, and the type of substrate. At greater depths in the lake lies the profundal zone, where the organisms live in perpetual darkness and depend on organic materials arriving from above, especially the continuous rain of phytoplankton cells sinking out of the photic zone (see Chapter 5).

The phytoplankton in most lakes contains dozens if not hundreds of species, and these fall into four major groups. First and foremost are the diatoms, with their highly ornamented walls of silica glass. These algal cells can be observed and identified with a standard microscope, and Forel appreciated their importance in the Lake Geneva food web, writing in 1904:

> a diatom, is eaten by a rotifer, which is eaten by a copepod, which is eaten by a cladoceran, which is eaten by a whitefish, which is eaten by a pike, which is eaten by an otter or by a human.

Glass is a heavy substance, and diatoms are therefore at the mercy of gravity, many of them sinking out of the photic zone and accumulating in the sediments. The best time of year for their growth is during spring and autumn, when the full mixing of the lake keeps these heavy cells up in suspension, and when there is adequate sunlight and nutrients for their photosynthetic production. There are long-term diatom records for many lakes of the world, including Lake Windermere, that show the regular rise and fall of diatoms every year. Their seasonal collapse is typically due to the onset of stratification, and the cessation of full mixing. Their rapid demise through sinking may also be hastened by grazing by zooplankton, or by parasitic attack.

Non-swimming green algae are also common in the phytoplankton, and vary enormously in size and shape. The smallest are less than 2 or 3μm in size and are referred to as 'picoeukaryotes'. These can be present in high abundance, but usually require DNA techniques to identify. At the other extreme, some species of green algae form large colonies of cells. For example, a beautiful lace-like species found only at Lake Biwa, Japan, *Pediastrum biwae* variety *triangulatum*, has a colony diameter of almost a tenth of a millimetre. A common species found in many lakes of the world, the gelatinous colonies of *Sphaerocystis schroeteri*, is of a similar or larger size. These dimensions make the particles too big for zooplankton to readily feed on, and large colony size is an effective defence against grazers.

The third group of phytoplankton comprises photosynthetic species that can swim: 'phytoflagellates' that belong to many algal phyla and encompass a great variety of species, both in terms of ecology, and pigmentation. These motile cells propel their way through the viscous liquid environment with flagella, some with one large and one very small flagellum such as the golden-brown species *Dinobryon divergens* that produces tree-like colonies of beating cells (Figure 20), others with two equal flagella such as

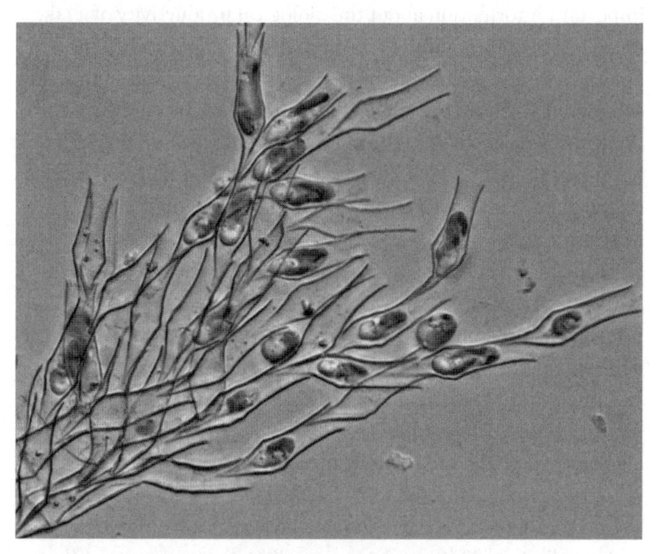

20. The colonial phytoflagellate, *Dinobryon divergens*. The individual cells are 10–15μm in length and are mixotrophic: they feed on bacteria as well as capture sunlight for photosynthesis.

the green alga *Chlamydomonas*. A mammoth among these motile phytoplankton is the brown-coloured dinoflagellate, *Ceratium hirundinella*, found in many lakes throughout the world; the cells are up to a quarter of a millimetre in length (250μm) and are able to swim up and down in the epilimnion each day.

The fourth group is the cyanobacteria, also known by their earlier name 'blue-green algae', with their distinctive combination of green chlorophyll and blue protein pigments. These include picoplankton species (picocyanobacteria) but also large colonial forms such as *Microcystis aeruginosa*. Cyanobacteria in general prefer warm temperatures and their large colonies are especially abundant in late summer and autumn, when they may form dense blooms and create major water quality problems.

The total amount and composition of the phytoplankton provide important information about the biological productivity of a lake as well as its water quality. The most rigorous approach is to analyse a sample using an inverted microscope; a glass-bottomed cylinder of lake water is allowed to sediment and is then examined though inverted lenses in the microscope that look up through the bottom of the cylinder to view the phytoplankton that have settled on the glass. This is a time consuming analysis, and it also requires a high level of skill by the microscopist to differentiate algal cells from detritus, and to identify the species.

Another, complementary approach is to measure the amount of chlorophyll *a*, a pigment that is found in all phytoplankton, including cyanobacteria. As a further estimate of the abundance of phytoplankton, as well as a guide to what groups are present, algal accessory pigments can be measured by high pressure liquid chromatography. In most samples, these analyses will show the presence of light-capturing pigments such as fucoxanthin in diatoms and peridinin in dinoflagellates, as well as diverse pigments that protect the cells against photodamage by bright light, such as lutein in green algae and echinenone in cyanobacteria.

This brings us back to Hutchinson's paradox: how can so many different species co-exist in the microscopic world of plankton? He considered several possible explanations, including the idea that the community might be in a state of disequilibrium; the lake environment is continuously changing, and so one species that is the winner today will be less favoured tomorrow, leading to a mixture of species and insufficient time to completely exclude any losers before their optimal conditions return. This idea has been further developed in the era of genomic analyses, which has revealed how the planktonic diversity of the lake microbiome is so much greater than even Hutchinson could have envisaged. Microbiologists talk about 'the rare biosphere' where most microbial species, including phytoplankton, are in low abundance and growing slowly or not at all for much of the time, while the most abundant species are subject to the heaviest losses by viruses and grazers. At the scale of a lake, even a residual population of a few cells per millilitre translates into a huge lake-wide population, which provides a hedge against extinction, and an inoculum to seize the day during the next round of favourable growth conditions.

Chapter 5
Food chains to fish

> The small and the weak are prey for the large and the strong; they in turn are devoured by the larger and stronger, or, if they escape, they will not avoid the microbes of decomposition that all organisms, directly or indirectly, are subject to.
>
> F. A. Forel

In his autobiography, François Forel recounts how one of his most exciting moments in lake science was discovering animal life at the bottom of Lake Geneva. The first sign came from his analysis of ripples in the sediments, just offshore from the family home at Morges. He had placed a sample of this bottom material under the microscope to determine its composition when suddenly a worm-like creature came into view, thrashing among the mineral particles. He was startled by this unexpected appearance of something so obviously alive, and he immediately began to wonder if the sediments of Lake Geneva could be inhabited by such animals to its greatest depths. If this were the case, then 'the profundal zone is not a desert; there is an abyssal society'.

That night, Forel constructed a dredge to sample deeper sediments in Lake Geneva, and in a set of studies that extended from the next day to the rest of his career, he discovered that a great variety of invertebrate animals occurred in the profundal zone (Figure 21), to the very bottom of the lake at around 300m depth. These bottom-dwelling or so-called 'benthic' communities of animals are dependent on the downward rain of organic materials, especially plankton, sinking into the profundal zone from the overlying waters. Forel called this continuous supply from above 'leftovers from the tables of others', and he realized that the benthic communities 'collect everything that falls to the bottom'. In turn, they are a food supply for other animals, including bottom-feeding fish, while bacterial decomposition in the sediments recycles the organic matter from all sources back into dissolved nutrients. Some four decades later, American ecologist Raymond L. Lindeman noted 'the brilliant exposition of Forel' on food webs, bacterial decomposition, and recycling. In his doctoral studies at Cedar Bog Lake, Minnesota, Lindeman built upon and greatly extended these ideas in a quantitative way to produce the 'trophic-dynamic' concept of energy and carbon flow, with bacteria and detritus in the bottom sediments forming the hub that linked all components of the food web.

Higher up in the water column, the life support system (protists and bacteria) described in Chapter 4 provides the carbon and energy for not only the benthic communities, but also for the animal residents of the pelagic or limnetic zone (Figure 21), the

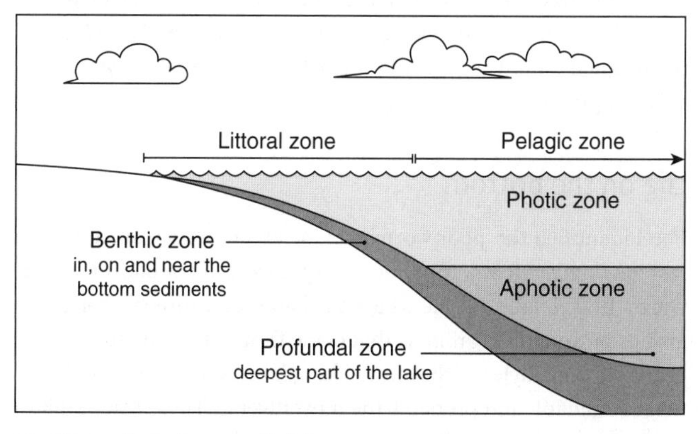

21. The ecological zones of a lake.

zooplankton. The largest of these can be readily discerned in a sample of lake water as minute, swimming individuals, 0.2–2mm in size, often moving in discontinuous jumps rather than smoothly gliding in a single direction. These are fed upon by fish that are themselves the prey for others, including larger fish, as well as birds and humans. A great variety of technologies, from high frequency acoustics and satellite telemetry to lipid, isotopic, and genetic analysis, are now combined with the usual methods of observation to better understand the nature of pelagic and benthic food webs and the coupling between them.

Recent studies of aquatic food webs have drawn attention to the importance for many lakes of the role of outside inputs from the surrounding catchment. These are derived from terrestrial plants and soils, and, in this way, they provide subsidies of carbon and energy to meet the needs of the aquatic animals (Figure 16). An additional external influence on lake food webs, but coming from well beyond the limits of the catchment, is the arrival of invasive species. Some of these plants and animals are introduced intentionally to 'improve' the ecosystem. Others accidently make their way into the lake, generally, and with increasing frequency, as a result of human activities like boating and fishing. In many cases, these invaders can severely disrupt the original food web, often to the detriment of the ecosystem services provided by the lake.

Life on the bottom

Forel identified the 'poor worm' moving about in such an agitated fashion on his microscope slide as a species of nematode, or round worm. In fact, lake sediments are the habitat for three important groups of worm-like animals that come from totally different branches (i.e. phyla) of the animal kingdom. Nematodes are the most abundant, and probably the most diverse. These thread-like invertebrates are typically 0.2–2mm in length, and they can occur in concentrations of up to a million per square metre of the lake floor. The smallest individuals are the most abundant, and most live in the upper few millimetres of the sediment. Some 2,000 freshwater species have been described, but the group is not well studied and it is estimated that many thousand more await discovery. They also vary greatly in their feeding strategies; some feed on aquatic plants, protists, or micro-invertebrates, some are larger animal parasites, while others, the dominants in lake sediments, feed on organic particles (detritus), bacteria, or microscopic fungi.

The second group is oligochaetes or segmented worms. Some of these move and burrow through the lake mud and silt, and certain species such as *Peloscolex variegatum* are restricted to well-oxygenated sediments. A major subgroup, the sludge worms, produces a tube of mucilage and particles that is embedded vertically in the sediment; the animals occupy the tube with their heads buried in the sediment for feeding, while their tails protrude through the tube into the overlying water and wiggle about to draw in oxygen. These oligochaetes can be brightly coloured with a red blood pigment that aids their survival in low oxygen environments. Two well-known representatives of this group are *Tubifex tubifex* and *Limnodrilus hoffmeisteri*, which are often found in sediments affected by organic pollution and are indicators of poor water quality.

The third group of worm-like animals in lake sediments is actually the larval form of insects, specifically flies (dipterans). The most common larvae are those of non-biting midges, chironomids (chi- pronounced 'kye' as in kite), and more than 5,000 species are known. These larvae are sought-after food items for benthic fish and other animals, and they can be found in large densities—up to tens of thousands per square metre of lake sediment. Given their much larger body size relative to nematodes, chironomids often dominate the total biomass of benthic communities. Many species have feeding tubes and use their body movement to generate currents that draw in oxygenated water; this burrowing activity as 'ecosystem engineers' can greatly change the oxygen conditions and biogeochemistry of the sediments. Like most other groups of benthic animals, the diversity of species and abundance is often greatest in the littoral zone (Figure 21) because of the diversity of substrates, the presence of plant and algal detritus, and the inputs of organic matter from the catchment. However, in moderate to large size lakes, the area of the littoral zone is small relative to that of the profundal zone, and the total lake biomass of these deep communities may be greater.

Chironomids figure prominently in the history of lake science because they were a favorite research topic for the distinguished zoologist August Thienemann, who was director of the hydrobiological laboratory at Plön, Germany. One of his international colleagues, the aquatic botanist Einar Naumann, had established a field station for the limnological institute at Lund, Sweden, and developed a classification scheme for lakes based on their algal concentrations. He referred to this as the lake's trophic state, from the Greek 'trophikos', meaning nourishment, and he classified waters into the categories of 'oligotrophic' (i.e. with clear water and low phytoplankton abundance) and 'eutrophic' (i.e. rich in phytoplankton). Thienemann adopted Naumann's classification, in common use today, and he showed how totally different assemblages of chironomids occur in these two trophic states; for example, the genus *Tanytarsus* is common in oligotrophic waters while *Chironomus* occurs in eutrophic waters that are low in oxygen. The two lake scientists joined forces in 1922 to found the International Limnological Society (SIL) at an inaugural meeting in Kiel, Germany.

Molluscs are another group of animals that occur in high abundance on the bottom of lakes, with two subgroups: gastropods (i.e. snails) and bivalves (i.e. clams and mussels). Certain fish species prey upon snails, which may find refuge in the littoral zone, on and among the water plants. In this habitat, they graze on detritus and on the biofilms of algae (i.e. periphyton) that coat the plants and the bottom of the littoral zone. Clams are renowned for their long life cycles, with a lifespan of several decades, and for some unionid species, for their impressive dispersal mechanism. These clams have gill membranes that extend outside the shell, in the shape of a small fish (sometimes even with pigmentation suggesting an eye); the membrane undulates with the pumping beat of the clam, and lures in predatory fish. Once the fish bites, the membrane breaks open and releases propagules called 'glochidia'. These latch onto the gills of the unsuspecting fish, and grow as small clams, finally dropping off once they are heavy enough, into the sediments, far away from the site of the parent.

A third group that often constitutes a large fraction of the benthic animal biomass is amphipods, also known as freshwater shrimps or scuds, and they are crustacean scavengers that mostly feed on detritus. Almost 2,000 species live in freshwater, and although a single species may appear to dominate the community, DNA analyses are showing that there are likely to be many hidden or 'cryptic' species that may be morphologically identical but genetically distinct. The greatest adaptive radiation has been observed in Lake Baikal; 260 endemic species have been recognized to date, with an additional eighty subspecies, and hundreds more are likely awaiting detection. Forel observed that an amphipod, 'the blind shrimp', *Niphargus forelii*, was common on sediments throughout the profundal zone of Lake Geneva (Figure 22). The species was subsequently driven to extinction in this lake, but it is still found today in other Swiss, German, and Italian deep alpine waterbodies. In the North American Great Lakes, the amphipod *Diporeia* can account for 50 per cent of the total biomass of benthic invertebrates, and in Lake Biwa, Japan, the endemic species *Jesogammarus annandalei*, can achieve densities up to $63{,}000/m^2$.

Apart from worms, molluscs, and amphipods, many other animals live in the benthic habitat of lakes, but generally at much lower biomass. These include small species such as rotifers and water mites, and crustaceans in addition to amphipods including

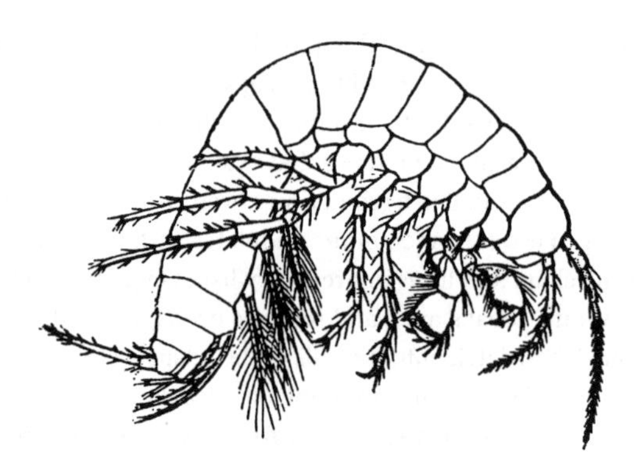

22. The blind shrimp (*Niphargus forelii*) of Lake Geneva.

ostracods (seed shrimps), harpacticoid copepods, and certain cladocerans, especially chydorids. Crayfish occur in a variety of freshwater benthic habitats and often have regional names. For example, they are known by the Maori name koura (*Paranephrops*) in New Zealand lakes, the aboriginal name 'yabby' (*Cherax*) in Australia, and as crawfish (*Procambarus clarkii*) in the southern US, where they are farmed as a sought-after ingredient of Cajun cooking. Crayfish are omnivores, and they mostly keep to the littoral zone where they feed on plant and detrital material, snails, chironomids, and mayflies; in turn they are eaten by fish and birds. Freshwater sponges occur in many lakes, and the polyp stage of freshwater jellyfish may also be found in the benthos, attached to underwater plants and other substrates. As Forel discovered to his wonder, the bottom of lakes, even the deepest of lakes, is most certainly 'not a desert' but rather a zone of biological richness and animal productivity.

Planktonic webs of interaction

In the open waters of the lake, three groups of zooplankton play a leading role in the transfer of carbon and energy from the base of the food web (i.e. phytoplankton and bacteria) to pelagic fish: rotifers, cladocerans, and copepods. The first of these groups is placed in its own branch of animal life, Phylum Rotifera, and is so named because of the rotating, wheel-like appearance of a double crown ('corona') of thread-like cilia that propels it through the water and directs a stream of food particles towards its mouth. These animals were first discovered in a drop of pond water by pioneer microscopist Antonie van Leeuwenhoek, who named them 'wheel animalcules'. They are rare in the sea, but in terms of numbers they are often the most abundant zooplankton in freshwaters. In thermokarst lakes across the northern tundra, for example, they can achieve densities up to 1,500 animals per litre. Rotifers are typically small in size (< 0.2mm), with fast generation times, often just a few days. Most feed on micro-algae, other protists, and bacteria, but there are also carnivorous species, notably in the genus *Asplanchna*. These animals in turn are eaten by copepods and young fish.

The second zooplankton group is cladocerans, which are crustaceans in the size range 0.5–2mm. There are eighty-six genera, of which only four occur in sea. Three of the most common planktonic genera are *Daphnia* (also known as water fleas, although they are not at all parasitic like fleas), *Bosmina*, and *Holopedium*, the latter distinguished by its giant, jelly-containing helmet that covers the head and that may be a defence against predation. Generally, this group is the favourite prey of smaller 'planktivorous' fish. The cladoceran group known as 'chydorids' (family Chydoridae) are found especially in the littoral zone associated with aquatic plants and bottom sediments. The body of cladocerans is covered in an exoskeleton of chitin that is shed as they grow and moult, this process occurring more than twenty times in some species. In some very clear lakes, their carapaces may contain dark melanin pigmentation (as in Figure 23) that acts as a sunscreen in protecting the animal and its eggs against the damaging effects of UV radiation.

Cladocerans have multiple pairs of appendages, with each pair specialized for particular functions. The most prominent

23. Photomicrograph of the zooplankton species *Daphnia umbra* from a lake in Finland. Each animal is around 2mm in length.

appendages are large antennae that are used as paddles for swimming and pulling the animals through the water, which is a viscous medium at the scale of the animal. The legs (four to six pairs) have fine hair ('setae'), with even finer hairs called 'setules' that filter particles from the water, including by electrostatic interactions. This food consists of algae, other protists, bacteria, and detritus, and the collected material is taste-tested by another set of antennae, ground up by other appendages (the mandibles), and worked with mucous into ball of matter ('bolus') that is then moved into the mouth or rejected.

Like rotifers, cladocerans can achieve spectacular population growth over short periods of time as a result of 'parthenogenesis' (Figure 24), and it is common to find temporary ponds and pools where clouds of *Daphnia* seem to appear suddenly from nowhere. Most of the populations are females, and these produce eggs asexually (i.e. without fertilization) that develop in the brood

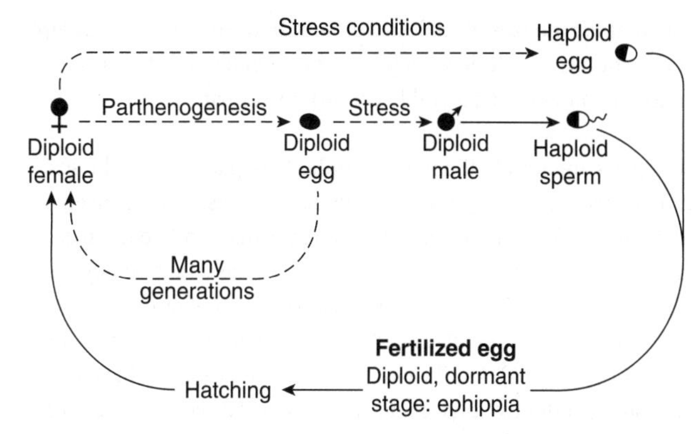

24. Asexual (parthenogenesis) and sexual reproduction by cladoceran zooplankton.

chambers, eventually released as free-swimming 'neonates'. Depending on the species and food conditions, a single mother may carry anything from one egg to more than 200, and in warm water the embryo development time can be as little as two days.

This asexual strategy is extremely effective for rapid growth in stable environments, but cladocerans like some rotifers have the advantage of a sexual mode of reproduction when conditions deteriorate (Figure 24). This could be due to a physical stress such as extreme temperatures, or biological stresses such as crowding and food shortages. At this time the females produce haploid eggs with a single copy of the chromosomes in each egg, and diploid eggs (double copy of each chromosome) that hatch into males. These males then mate with the females and fertilize their haploid eggs to produce a diploid zygote. In many species this is retained with a modified part of the carapace, and is released during moulting as a darkly coloured, encased resting egg called an 'ephippium'. These are highly resistant to extremes such as drying and freezing, and may be important for dispersal between waterbodies in the wind and via bird feathers. They can remain dormant for months, decades, even hundreds of years, for example at the bottom of a desiccated pond, and then hatch into asexual, diploid females once favorable conditions return.

Copepods are among the most abundant zooplankton in the sea, and are also commonly found in almost all lakes. In large deep lakes such as Lake Baikal and the North American Great Lakes, they account for most of the zooplankton biomass. Like cladocerans, copepods are crustaceans with an exoskeleton of chitin and multiple pairs of appendages adapted for swimming, feeding, or sensing (Figure 25). Unlike cladocerans, however, they have no asexual reproductive phase and the populations contain a mixture of males and females. After mating, the female produces eggs that hatch into larvae or 'nauplii', which then moult five or six times

25. Photomicrograph of the copepod *Aglaodiaptomus leptopus* from a lake in southern Quebec, Canada. The animal is 2.3mm long.

before becoming copepodid larvae ('copepodites'). These then undergo five more moults before becoming sexually mature adults. In warm water this entire life cycle may be completed within a week, but in cold lakes of the polar and alpine regions it may take one or more years. Copepods feed on phytoplankton and other protists, and are a lipid-rich food source for planktivorous fish. However, they are harder to catch than are the slower swimming Cladocera, which is why they often dominate in lakes where fish planktivory is strong.

Moving about

Rotifers, cladocerans, and copepods are all planktonic, that is their distribution is strongly affected by currents and mixing processes in the lake. However, they are also swimmers, and can regulate their depth in the water. For the smallest such as rotifers and copepods, this swimming ability is limited, but the larger zooplankton are able to swim over an impressive depth range during the twenty-four-hour 'diel' (i.e. light–dark) cycle. Forel first observed this himself when he rowed out onto Lake Geneva to sample zooplankton at night, and found that his net tow had captured 'myriads of Entomostraca [copepods] that had risen to the surface'. Subsequent work has shown that the cladocerans in Lake Geneva reside in the thermocline region and deep epilimnion during the day, and swim upwards by about 10m during the night, while cyclopoid copepods swim up by 60m, returning to the deep, dark, cold waters of the profundal zone during the day.

Even greater distances up and down the water column are achieved by larger animals. The opossum shrimp, *Mysis* (up to 25mm in length) lives on the bottom of lakes during the day and in Lake Tahoe it swims hundreds of metres up into the surface waters, although not on moon-lit nights. In Lake Baikal, one of the main zooplankton species is the endemic amphipod, *Macrohectopus branickii*, which grows up to 38mm in size. It can form dense swarms at 100–200m depth during the day, but the populations then disperse and rise to the upper waters during the night. These nocturnal migrations connect the pelagic surface waters with the profundal zone in lake ecosystems, and are thought to be an adaptation towards avoiding visual predators, especially pelagic fish, during the day, while accessing food in the surface waters under the cover of nightfall.

One of the most striking effects of fish on migration patterns is for the insect known as the 'phantom midge'. This produces mosquito-like larvae up to 2cm long, also called 'glass worms' because their bodies are transparent apart from a pair of air bags at each end that help them float in the water. In Europe there are two main species that differ in their diel migration behaviour. *Chaoborus flavicans* occurs mainly in fish-containing lakes and ponds, and during the day it remains at the bottom, feeding on animals in the sediments, including in anoxic waters or with its head dug into anoxic sediments where it appears to be capable of anaerobic metabolism based on malate; during the night it migrates to the surface to feed on zooplankton, especially copepods. This migration pattern is most pronounced in the presence of fish, which it appears to be able to detect through chemical signals (fishy-smelling 'kairomones'). The second species, *Chaoborus obscuripes* avoids fish-containing waters, and remains throughout the twenty-four-hour cycle in the near-surface zone, where it is out of the reach of benthic predators such as dragonfly larvae.

Although certain fish species remain within specific zones of the lake, there are others that swim among zones and access multiple habitats. For example, in Lake Superior, the largest freshwater lake by area in the world (82,100km^2; maximum depth 406m), the Cisco (*Coregonus artedi*) is a fish species that is mostly planktivorous, living in the offshore pelagic zone. However, in late autumn it moves inshore to spawn, and its lipid-rich eggs provide an energy subsidy to the food web of the littoral zone, contributing 34 per cent of the energy needs of Lake whitefish (*Coregonus clupeaformis*) that feed mostly on benthic prey such as amphipods in these inshore shallow waters. This type of fish migration means that the different parts of the lake ecosystem are ecologically connected.

For many fish species, moving between habitats extends all the way to the ocean. Anadromous fish migrate out of the lake and swim to the sea each year, and although this movement comes at considerable energetic cost, it has the advantage of access to rich marine food sources, while allowing the young to be raised in the freshwater environment with less exposure to predators. One such example is Arctic char (*Salvelinus alpinus*), a species that is found in deep cold lakes in Great Britain as well as Lake Geneva and elsewhere in Europe. It is the northernmost freshwater fish, and is found all the way to Lake A at latitude 83°N in the Canadian High Arctic. During their freshwater residence, Arctic char feed on benthic invertebrates as well as plankton, small fish, and insects at the water surface; in the sea these animals feed on other fish and amphipods. Genetic markers along with acoustic tags inserted into captured and released fish are now being used to identify distinct populations of this species, and the origin of migrating stocks.

With the converse migration pattern, catadromous fish live in freshwater and spawn in the sea. One example is the European eel, *Anguilla anguilla*, which is most abundant in rivers but also commonly found in natural and artificial lakes. Satellite tagging has shown that adult European eels migrate 5,000km or more to spawn in the Sargasso Sea. At swimming speeds in the range 10–30km per day, this is a long process, taking up to a year and with severe losses by predation. The larvae then return to European waters by the Gulf Stream and North Atlantic Drift.

You are what you eat?

Diet certainly has a major role to play in the nutritional state of animals in the food web of lakes, but changes can also take place via the specific needs and physiology of the species concerned. Take something as simple as the ratio of elements. The famous American oceanographer Alfred C. Redfield established that phytoplankton in the sea typically have a ratio of 106 atoms of carbon to sixteen atoms of nitrogen to one atom of phosphorus (or in terms of weight, 41 grams (g) of carbon to 7g of nitrogen and 1g of phosphorus). He also noted that the same N:P ratio occurred in the regenerated nutrients in the deep ocean. In lakes, the phytoplankton often has a carbon to nitrogen to phosphorus composition that is similar to or slightly above this 'Redfield ratio', but higher up the food web this ratio can differ greatly among animals. The nitrogen to phosphorus ratio for copepods is typically 14g of nitrogen to 1g of phosphorus (i.e. 14:1) or higher, but for cladocerans, even in the same lake, it is often half of that, around 7:1. This strikingly lower ratio in cladocerans has been attributed to their greater quantity of cellular organelles for protein production, specifically ribosomes that contain RNA, a biomolecule that is rich in phosphorus. This allows faster growth rates, but also results in a high biological demand for phosphorus, hence the lower nitrogen to phosphorus ratio. The analysis of nutrient ratios in lakes, 'ecological stoichiometry', has generated important insights and questions in lake science, including about the divergent needs of different animal groups for phosphorus and effects on nutrient recycling.

Food quantity and its rate of supply are important to all animals in the food web, and in general terms, the animal or 'secondary productivity' of lakes increases with increasing phytoplankton and its associated photosynthetic production of biomass or 'primary productivity'. But it is not just about quantity. The nutritional quality of that food can vary greatly, for example in its composition of fatty materials (lipids), and this profoundly affects animal health, reproductive success, and survival. Some of these lipid molecules are brightly coloured and are derived from algal pigments that are transferred to the animals that feed upon them (e.g. the copepod in Figure 25 is bright orange in colour due to the carotenoid astaxanthin). For certain zooplankton, such as in clear alpine lakes, this pigmentation may be primarily a defence against UV radiation, but for other species it appears to be a way to maintain high energy lipid reserves over winter for growth and reproduction the next spring.

Certain lipid molecules called 'polyunsaturated fatty acids' (PUFAs), such as the omega-3 PUFA, eicosapentaenoic acid (EPA), are especially important for the production of hormones that regulate bodily functions such as brain development, vision, and cardiovascular metabolism (including in humans). Most aquatic animals cannot produce PUFAs but must obtain them from their diet, ultimately from the algae that produce them and the consumers of those algae. There is great interest by lake scientists in tracking EPA and related PUFAs through the food web because this provides information on the feeding relationships and health of the aquatic ecosystem. There may even be a transfer of PUFAs to terrestrial species (which tend to be poorer in PUFAs than aquatic organisms) via aquatic insects such as chironomids and dragonflies that are eaten by birds. Lipids are also relevant to understanding the impacts of chemical pollution. Most organic contaminants (e.g. pesticides) are soluble in lipid, and therefore the transfer of lipids through the food web also results in the concentration (i.e. bioamplification) of pollutants in the animals at higher positions (or trophic levels) in the food chain.

One of the most powerful approaches towards analysing food webs is based on naturally occurring isotopes: atoms that have the same number of protons, and are therefore of the same element, but that differ in the number of their neutrons. Nitrogen in the atmosphere, for example, is mostly composed of atoms with seven protons and seven neutrons, denoted 'nitrogen-14' (or, ^{14}N), but a small percentage of it (0.3663 per cent) is composed of nitrogen atoms with an additional neutron, making it nitrogen-15 (^{15}N). When nitrogen is taken up by animals, the ^{15}N is retained to a slightly greater extent than ^{14}N, and this enrichment effect continues step-by-step up the food chain.

These differences based on a single neutron may seem small, but with sensitive mass spectrometers, even minute shifts in ^{15}N enrichment can be accurately detected. In the pelagic zone of Lake Baikal, for example, ^{15}N is enriched by about 3.3ppt, with each step in the food web (Figure 26). Large endemic diatoms in the phytoplankton such as *Aulacoseira baikalensis* take up inorganic

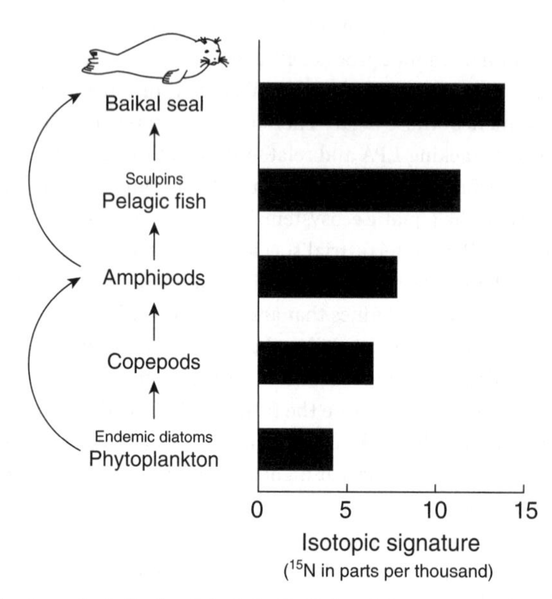

26. The pelagic food web in Lake Baikal and the corresponding increase in nitrogen-15 ($\delta^{15}N$) at each trophic level.

nitrogen, and their delta ^{15}N (denoted as $\delta^{15}N$ and defined as the difference in the $^{15}N/^{14}N$ ratio relative to the atmosphere) is around 4ppt. The diatoms are eaten by copepods and eventually the nitrogen passes all the way up the food chain, with a final $\delta^{15}N$ value of 14ppt in the seals. There are other sources of variation, but the approach provides a valuable guide to who is eating who in the lake food web, and for omnivorous animals, the relative importance of different food items in their diet. The natural isotopic ratio $^{13}C/^{12}C$ is also used in this way to measure dietary source, and, in marine studies, the sulfur isotopic ratio $^{34}S/^{32}S$ provides a similar tracer. Isotopic fractionation also takes place when water evaporates and passes from liquid to gas, and the isotopic ratios $^{2}H/^{1}H$ and $^{18}O/^{16}O$ in water molecules are used in hydrology to determine the evaporation versus precipitation balance of lakes.

Invaders at the lake

In the late 19th century, Forel observed with alarm the arrival of the Canadian water weed *Elodea canadensis* in Lake Geneva, and its 'exuberant and frightening expansion' throughout the lake. This invasive species had been introduced intentionally in local ponds and streams to improve the fish habitat, and in an unfortunate history repeated in many parts of the world, it soon entered and expanded throughout the littoral environment of the lake. This species and other members of the aquatic plant family Hydrocharitaceae invaded New Zealand lakes in the mid-20th century, forming underwater forests up to 6m tall that greatly altered littoral habitats and interfered with electricity production via their proliferation in hydro-reservoirs. Other species such as the *Myriophyllum spicatum*, the Eurasian water milfoil, is creating problems in drinking water supplies (including at Lake St-Charles, Quebec), and the South American water hyacinth *Eichhornia* is an aggressive floating plant that covers and chokes aquatic habitats in the southern USA, Asia, and Africa, including inshore sites at Lake Victoria.

The arrival of an invasive animal species in a lake can have a massive effect that begins at one level and then propagates throughout the entire food web. A classic example of this 'trophic cascade' was observed in Flathead Lake, a large (500km^2), deep (116m) lake in Montana, USA. Over the period 1968 to 1975, the opossum shrimp, *Mysis diluviana* (closely related to the European species, *Mysis relicta)* was introduced into three headwater lakes to improve the salmon fishery. By 1981, the shrimp had drifted downstream to appear in Flathead Lake, where by the late 1980s it had undergone an explosive increase in numbers (Figure 27). Within a few years after the arrival of the mysids at Flathead Lake, there was a collapse of cladocerans and copepods in the zooplankton due to overconsumption by the massive shrimp populations. A 'top-down' food web effect followed, with large increases in phytoplankton biomass and changes in its community composition because of decreased grazing by zooplankton.

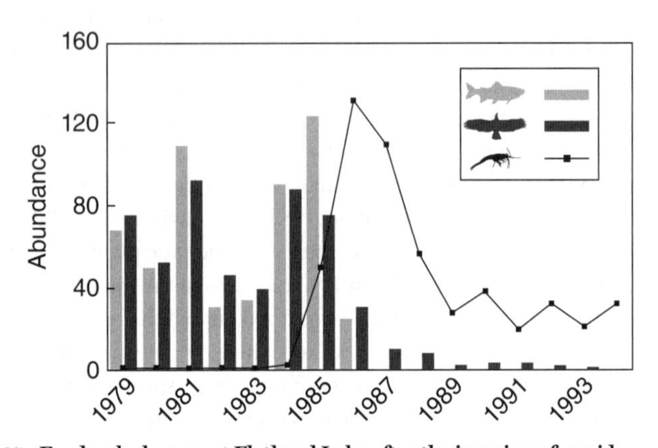

27. Food web changes at Flathead Lake after the invasion of mysid shrimps. The numbers of Kokanee salmon (the plotted values should be multiplied by a hundred) and Bald eagles (multiply by seven) were measured at an upstream salmon spawning site, while the shrimp abundance (multiply by 1,000) was the number of mysids in the water column of the lake per square metre.

The Kokanee salmon, also an introduced species in Flathead Lake, became deprived of their zooplankton food source, and were unable to feed on the mysids, which migrated up into the pelagic zone only at night when the salmon were unable to see them. This unforeseen fish avoidance behaviour has made mysid stocking an especially poor choice for enhancing salmon production. In the Flathead Lake watershed, sport catches of salmon plunged from more than 100,000 through 1985 to zero in 1988 and beyond. Bald eagles, which congregated on the main spawning stream to feed on Kokanee, went from more than 600 in the early 1980s to near-zero a decade later (Figure 27). An additional food web effect was that the mysids became the major food source for another introduced species, bottom-feeding Lake trout (*Salvelinus namaycush*), and the current rise of this species is driving the native Bull trout (*Salvelinus confluentus*) towards extinction.

Invasive species that are the most successful and do the most damage once they enter a lake have a number of features in common: fast growth rates, broad tolerances, the capacity to thrive under high population densities, and an ability to disperse and colonize that is enhanced by human activities. Zebra mussels (*Dreissena polymorpha*) get top marks in each of these categories, and they have proven to be a troublesome invader in many parts of the world. Their native habitat is in the Caspian Sea region, but with the construction of canal systems throughout Europe in the 18th and 19th centuries, they soon became well dispersed, arriving in Great Britain by 1824. They were first discovered in the North American Great Lakes in 1988, and are believed to have arrived, like many other invasive species, in the ballast waters of cargo ships. By 1990 they were well established throughout the Great Lakes, and have now moved into the Mississippi River basin. A related species, the Quagga mussel (*Dreissena bugensis*) also invaded the Great Lakes at around the same time and has created additional problems by colonizing soft sediments and deeper water than the Zebra mussels.

A single Zebra mussel can produce up to one million eggs over the course of a spawning season, and these hatch into readily dispersed larvae ('veligers'), that are free-swimming for up to a month. The adults can achieve densities up to hundreds of thousands per square metre, and their prolific growth within water pipes has been a serious problem for the cooling systems of nuclear and thermal power stations, and for the intake pipes of drinking water plants. A single Zebra mussel can filter a litre a day, and they have the capacity to completely strip the water of bacteria and protists. In Lake Erie, the water clarity doubled and diatoms declined by 80–90 per cent soon after the invasion of Zebra mussels, with a concomitant decline in zooplankton, and potential impacts on planktivorous fish. The invasion of this species can shift a lake from dominance of the pelagic to the benthic food web, but at the expense of native unionid clams on the bottom that can become smothered in Zebra mussels. Their efficient filtering capacity may also cause a regime shift in primary producers, from turbid waters with high concentrations of phytoplankton to a clearer lake ecosystem state in which benthic water plants dominate.

The problem of invasive species is now compounded by global climate change. This is weakening the competitive abilities of native plants, animals, and microbes at the upper end of their thermal range, and it is opening new habitat opportunities for species from the warm, temperate, and tropical regions to expand into previously cold, unfavourable lakes. Conservation regions such as national and regional parks are now more important than ever in protecting lake food webs from additional stressors, and in lessening the pressure of invasive species that inevitably accompany land development and the associated expansion of transport routes.

Chapter 6
Extreme lakes

> The composition of lake water that we recognize at the surface, is it maintained at all depths, or instead does it vary, and in what manner?
>
> F. A. Forel

The familiar, upper waters of a lake are often a poor guide to what lies well beneath the surface, and in some lakes the variations with depth are extreme. One of the most striking examples is Lake Vanda in the McMurdo Dry Valleys region of Antarctica, where thick ice overlies the water throughout the year and prevents mixing by the wind. When the first scientists drilled a hole in the ice and lowered their thermistor probe down through the underlying water column, they were surprised to discover that temperatures rose with increasing depth, finally reaching 26°C at the bottom. This temperature inversion of warm water underlying cold was possible because of a strong gradient in salt concentration: while the surface waters of Lake Vanda are fresh and derived from pure glacial ice, its bottom waters are up to three times the salinity of seawater. After a period of lively debate, this unexpected warmth was finally explained as the cumulative effect of sunshine: year after year, century after century, solar radiation in summer has passed through the clear ice and freshwater, and has gradually warmed the dense, salty bottom layer to the unlikely temperatures observed today.

Extreme lakes are waterbodies that have unusual physical, chemical, and biological features, and they are of great scientific interest. Salt water lakes occur in many parts of the world and are often highly productive, with a simplified food chain that supports large flocks of resident or migrating birds. Polar and alpine lakes are strongly affected by snow and ice, and are therefore sensitive to small changes in temperature across the freezing–melting threshold of water. These high latitude and high altitude ecosystems are global sentinels of climate change in the past and present, as well as models for wider understanding of lake microbiology and biogeochemistry. Other extremes among the world's lakes include acid and alkaline waters, geothermal hot water lakes, and waterbodies that periodically erupt, disgorging their liquid and gaseous contents, and creating danger to anyone in the vicinity.

In the most severe lake conditions, only the hardiest of 'extreme-loving' microbes can survive and grow. These 'extremophiles' include halophiles that prefer the highest salinities, psychrophiles (from the Greek 'psukhrós' meaning cold or frozen) that are adapted to perennially cold water, and acidophiles that grow best under low pH. Biochemical and genomic research on these microbes is providing insights into the origins, evolution, and limits of life on Earth, and has yielded unique biomolecules of medical and biotechnological application.

Salt water lakes

One of the many distinguishing features of H_2O is its unusually high dielectric constant, meaning that it is a strongly polar solvent with positive and negative charges that can stabilize ions brought into solution. This dielectric property results from the asymmetrical electron cloud over the molecule, described in Chapter 3, and it gives liquid water the ability to leach minerals from rocks and soils as it passes through the ground, and to maintain these salts in solution, even at high concentrations.

Collectively, these dissolved minerals produce the salinity of the water, measured in terms of grams of salt or dissolved solids per litre. Since 1 litre (L) of water weighs 1,000g, salinity can be expressed as grams per thousand grams or ppt. Sea water is around 35ppt, and its salinity is mainly due to the positively charged ions sodium (Na^+), potassium (K^+), magnesium (Mg^{2+}), and calcium (Ca^{2+}), and the negatively charged ions chloride (Cl^-), sulfate (SO_4^{2-}), and carbonate (CO_3^{2-}).

These solutes, collectively called the 'major ions', conduct electrons, and therefore a simple way to track salinity is to measure the electrical conductance of the water between two electrodes set a known distance apart. Lake and ocean scientists now routinely take profiles of salinity and temperature with a CTD: a submersible instrument that records conductance, temperature, and depth many times per second as it is lowered on a rope or wire down the water column. Conductance is measured in Siemens (or microSiemens (µS), given the low salt concentrations in freshwater lakes), and adjusted to a standard temperature of 25°C to give specific conductivity in µS/cm.

All freshwater lakes contain dissolved minerals, with specific conductivities in the range 50–500µS/cm, while salt water lakes have values that can exceed sea water (about 50,000µS/cm), and are the habitats for extreme microbes, such as the halophilic, green algal flagellate *Dunaliella* and salt tolerant archaeons (Haloarchaea) that have biochemical strategies to contend with such a high level of salinity stress. Deep Lake in the Vestfold Hills of Antarctica is so saline (270ppt) that the water never freezes and one could row out into the middle of the lake in mid-winter while the surrounding landscape is deeply frozen; it would be best to stay out of the water, however, since the temperature of the liquid brine is around –18°C.

Saline waters account for a vast total area, and hold a number of records among the world's lakes. The largest lake in the world is the Caspian Sea, extending over 371,000km^2 with a maximum depth of 1,025m. It has a moderately high salinity (12ppt) that is derived from terrestrial rather than marine sources; in contrast, the Black Sea exchanges water with the Mediterranean Ocean and is therefore considered a marine system, not a lake. Like many saline lakes, the Caspian Sea is an ancient waterbody that occupies a tectonic basin, with numerous endemic species including a landlocked seal, *Pusa caspica*. One of the world's oldest lakes is the moderately saline (6ppt) Lake Issyk-Kul (meaning 'warm lake') that lies in the Tian Shan Mountain area of Kyrgyzstan. This large, deep lake (6,300km^2; maximum depth 702m) has an age rivalling that of Lake Baikal (around twenty-five million years) and supports a diverse fauna including endemic species. The Dead Sea at 400m below sea level is the lowest lake in the world and one of the saltiest: 342ppt, about ten times that of sea water. Salt lakes also occur at extreme high altitudes, including on the Tibetan Plateau and the altiplano of Bolivia and Peru. Despite all of these unusual features, saline lakes have been considered expendable because of their often remote location and salty, undrinkable water. However, their high value for migratory birds and importance to rare species has placed them on the front line of conservation battles in several parts of the world.

Conflicting views about the value of saline lakes were especially apparent in the long and ultimately successful battle to save Mono Lake, California. When Mark Twain visited the region in the early 1860s, he called the lake a 'solemn silent, sail-less sea' that lay in a 'hideous desert'. Yet like many saline waterbodies, Mono Lake is a place of stunning beauty, bountiful plankton, and immense flocks of water birds. The lake is considered a 'triple water' in that its saltiness is due to three components: carbonate (hence it is considered a soda lake), chloride, and sulfate. If you dip your hand into the water, it has a slippery, soapy feel to it; and then as the water dries rapidly in the desert sun, your hand will be left covered with a film of salt, like a thin, white glove.

28. Tufa towers at Mono Lake, California.

Underwater springs flow into Mono Lake, and when these cold freshwaters containing calcium meet the saline lake water, calcium carbonate precipitates out as the mineral calcite to produce pillars called tufa towers. This process is also aided by cyanobacterial films that coat the tufa towers, and their removal of CO_2 by photosynthesis shifts the equilibrium towards more carbonate precipitation. Many of these impressive towers are exposed around the edge of the lake by ancient and modern falling lake levels, and can be several metres in height (Figure 28).

Mono Lake lies on the eastern side of the Sierra Nevada mountain range of California, at the edge of a vast, high, desert region called the Great Basin. This area once contained extensive freshwater lakes, but these ancient waters have long since evaporated to leave behind salt pans and salt lakes. The largest of these residual waterbodies is Great Salt Lake, Utah (4,400km^2; maximum depth 14m), with 50–270ppt of salt, depending on its fluctuating water level. Great Salt Lake, Mono Lake, the Caspian Sea, and many other saline waterbodies, are 'endorheic', meaning they have no outflow. The large evaporative losses of Mono Lake water were offset by freshwater streams that recharged the lake each year with snowmelt from the Sierra Nevada Mountains. But water planners for the city of Los Angeles were casting their efforts far and wide to meet the needs of the rapidly growing population, and began to divert those Sierra Nevada meltwaters and channel them 560km through aqueducts to the city. The first major diversion began in 1941, and from that point onwards Mono Lake began to shrink in size and become increasingly saline. From the 1940s to the 1970s, the lake level fell by around 15m, and the salinity doubled from 40 to 80ppt. Water usage by the city of Los Angeles had already caused the complete drying up of Owens Lake, another endorheic lake in the region, and it seemed that Mono Lake was on a similar path to extinction.

The fortunes of Mono Lake began to reverse when a group of undergraduates from the University of California Davis and Stanford University teamed up to undertake a summer research programme on the ecology the lake in 1976. Their study drew attention to the highly productive food web of this lake, based on two hardy invertebrates: the alkali fly (*Ephydra hians*) that undergoes its larval and pupal stages in the water at the edge of the lake, in the past harvested as food by the Kuzedika Native Americans, and the brine shrimp (*Artemia monica*) that achieves a population of trillions each year in the lake. The brine shrimp feeds on a minute (less than 3μm) salt-tolerant green alga called *Picocystis*.

The student research at Mono Lake showed that the flies and shrimp were the food resources for huge numbers of migratory bird species that used the lake as an important stopover on their flyways each summer, including some 50,000 gulls, 80,000 phalaropes, more than a million eared grebes, and many other species. Most importantly, they showed that rising salinities would push the brine shrimp to extinction. This collapse of food would be compounded by the connection of islands to the shore because of the falling lake levels, which would expose nesting birds such as the California gull to coyotes and other predators. Members of this group, led by ornithologist David A. Gaines, went on to found the Mono Lake Committee that took the city of Los Angeles to court, while raising public awareness of the ecological value of the Mono Lake ecosystem and its dire trajectory. The Committee's tireless efforts over a marathon fifteen years of court battles eventually won legal and political support, culminating in the full re-instatement of inflows and a rise in lake levels. The lake is now a unique conservation park that draws many visitors each year as well as the prolific flocks of migrant water birds.

Polar and alpine lakes

Lakes at high latitudes and altitudes encompass a diverse range of ecosystem types, from the numerous floodplain lakes that wax and wane on Arctic river deltas, to the vast deep waters of Great Bear Lake in northern Canada (31,153km^2; maximum depth 446m), highly stratified waterbodies such as Lake Vanda, Antarctica, and deep, clear alpine lakes such as Lake Redon (2,240m above sea level; maximum depth 73m) in the Pyrenees, first studied by the eminent Catalan ecologist Ramon Margalef. Despite this variety of habitat types, polar and alpine lakes hold a number of features in common, including their remoteness from cities and the direct effects of pollution. This has made such lakes ideal sites to track the long range dispersal of heavy metals and organic contaminants. For example, at Lake Redon, organic contaminants such as DDT and PCBs have been detected in the water many decades after these compounds were banned from use in Europe, indicating their arrival from thousands of kilometres away, and the need to collaborate globally in the control of poisons in the biosphere. Other contaminants have been detected that arrive from regional sources (e.g. hexachlorocyclohexane pesticides that are used in agriculture in southern Europe). The remoteness of polar and alpine lakes has also been of great interest for biogeographical studies, and while there is some evidence of the cosmopolitan distribution of certain cold-tolerant microbes, for example some freshwater cyanobacteria, other studies of these isolated, island-like ecosystems have shown that there are regional assemblages, including cyanobacteria and single-cell eukaryotes (protists).

Another common feature of polar and alpine lakes is their intimate association with the cryosphere, the ensemble of snow- and ice-containing environments throughout the world. These lakes are covered by thick ice for much or even all of the year. This ice cap, often overlaid by snow, restricts the light available for primary production, and this effect is compounded in the polar regions by up to three months of continuous winter darkness. A major impact of climate warming on such lakes is the extension of the ice-free season, with earlier ice melt in spring and later freeze-up at the end of summer. This results in not only more light for photosynthesis, but also more upward mixing of nutrients for algal growth in the surface waters. However, under these open water conditions, the lake biota also has to contend with greater exposure to potentially harmful ultraviolet radiation, which may influence productivity and species composition.

One group of organisms that is pre-eminently successful in polar and alpine lakes is the 'psychrotolerant' cyanobacteria that can withstand extreme cold and even complete freeze-up, but grow more rapidly at warmer temperatures (hence they are not 'psychrophiles' that prefer the cold). These micro-organisms bind together sand and silt particles with their filaments and a mucilaginous glue made of sugary compounds to produce a thick biofilm or 'microbial mat' that coats the bottom of lakes, ponds, and streams. The mats are often bright pink or orange due to carotenoids that prevent UV-B damage from the bright incident sunshine, and they range in thickness from a fraction of a millimetre to several tens of centimetres. The most spectacular communities have been found at the bottom of permanently ice-covered Antarctic lakes, such as Lake Untersee, where they produce dome-like structures that bear a close resemblance to the earliest fossils ('stromatolites') on Earth.

In cold, ice-covered waters, cyanobacterial mats and films often occur in association with mosses, and have high concentrations of red and blue proteins called 'phycobiliproteins', which capture light for photosynthesis with high efficiency. These communities can dominate the primary production and biomass of polar and alpine lakes. Analyses of their microbiome composition is showing that although the dominants are cyanobacteria there are hundreds if not thousands of other microbes present: other bacteria as well as archaea, viruses, and eukaryotes such as diatoms and small invertebrate animals that make their home in nutrient-rich microhabitats provided by the cyanobacterial mat. It has been suggested that eukaryotic cells (protists) might have found refuge and thrived in these biofilms coating the base of melt pools on ice (as they do today on polar ice shelves and glaciers) during the 'Snowball Earth' glaciations that extended over most of the globe between about 720 and 635 million years ago.

Polar and alpine lakes are also useful model systems to study and better understand how elemental cycles function in the aquatic environment, and how lakes are affected by inputs from their surroundings. Some of the best natural laboratories for such studies are the permanently layered lakes in the polar regions, such as lakes Vanda, Fryxell, Hoare, Bonney, Joyce, and Miers in the McMurdo Dry Valleys region of Antarctica. These waterbodies are called 'meromictic lakes', meaning only partially mixed. Another lake district containing these saline, permanently layered waters occurs at the opposite pole, in the Canadian High Arctic. These lakes were first discovered by a military research expedition in 1969, and with tactical formality were named lakes A, B, C.... The lakes still retain these unexciting names from the Cold War era, but the letters belie their many unusual and intriguing properties.

Lake A (latitude 83°N; maximum depth 128m) lies in the national park called 'Quttinirpaaq', an Inuit word meaning 'Top of the World', and occupies a valley that some 5,000 years ago was filled with seawater as a fjord connected to the Arctic Ocean. With the melting of the Arctic ice sheets and the associated loss of massive pressure from above, the valley rose up out of the sea, cutting off the fjord as an isolated lagoon of Arctic Ocean water. Melting snow and glaciers produced dilute waters that discharged into the lake and floated on top of the dense, salty water beneath. The resultant layered nature of Lake A today can be seen by its salinity profile (Figure 29): low conductivity meltwaters occur beneath the ice as a surface layer, while at about 11m there is a sudden rise in salinity that continues down to the ancient seawater that was trapped in the valley thousands of years ago.

The salinity profile of Lake A provides clues to its geological history, while the temperature profile provides a record of more recent change. Its thermal profile shown in Figure 29 indicates some summer warming under the ice, likely associated with the

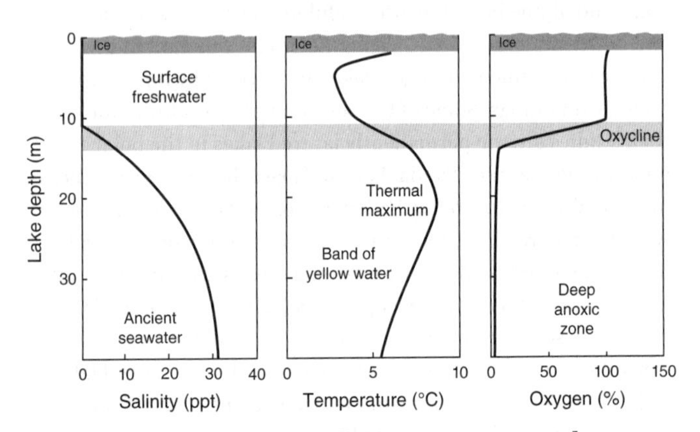

29. Layers of water of different salinities, temperatures, and oxygen in ice-capped Lake A in the Canadian High Arctic.

warm, low-salinity inflows, followed by a decrease in temperature in the deeper, underlying water. But then there is an unexpected rise to the thermal maximum at about 22m depth. As in Lake Vanda, Antarctica, the reason for this rise is the gradual warming by sunshine penetrating through the ice and water to this depth.

The surface waters of Lake A are fully saturated in oxygen, recharged in summer by ephemeral streams that flow from melting snowbanks, but deeper in the lake the oxygen values plunge across the oxygen gradient ('oxycline') to below the limit of detection, and these anoxic conditions extend to the bottom of the lake. The colour and smell of the water changes dramatically with depth and is further evidence of this layering. For example, a band of yellow-coloured water occurs at 28–30m depth, and is the result of green photosynthetic sulfur bacteria that capture light for energy and use H_2S (instead of H_2O used by plants) to reduce CO_2 to sugars, in the process producing yellow particles of elemental sulfur. The rotten-egg smell of H_2S is obvious in water samples brought up from 30m and below.

Molecular tools based on nucleic acids (DNA and RNA) offer a powerful set of approaches towards analysis of the layering of different microbial communities, and are now used routinely in lake studies. These are providing insights into microbial diversity and biogeochemical processes that were previously intractable because most members of the aquatic microbiome have not been brought into culture and cannot be distinguished under a microscope. Once DNA has been extracted and its nucleotides (the A, G, C, T alphabet of nucleic acids) sequenced, the genetic relationships can be explored by producing a tree-of-life diagram, where the distance between samples or species is a measure of relatedness. One of the strengths of this approach is that all data are shared in an international database (GenBank), which provides a huge (200 million records at present), ever improving reference source for these sequence comparisons.

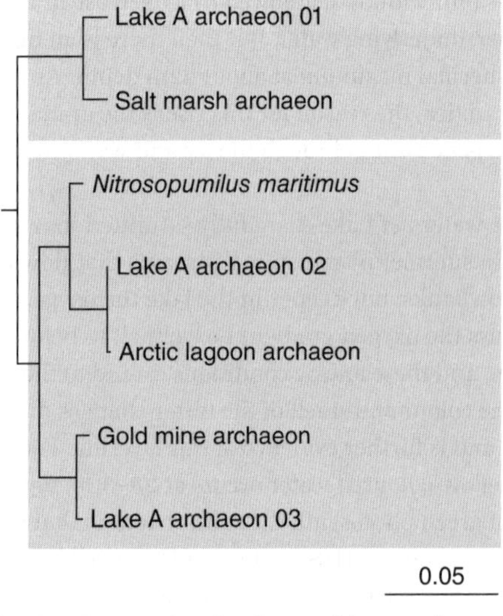

30. Tree showing the genetic relatedness of three archaeon strains in Lake A, and their close relationship to archaea from certain other habitats.

An example of this molecular approach applied to Lake A is shown in Figure 30. First, water was sampled from 10–12m depth where there was a sharp drop in oxygen (Figure 29). This is always an interesting place to look for new microbes because oxygen gradients usually contain a blended cocktail of oxidized and reduced chemicals, favouring a variety of microbial lifestyles. The DNA sequences of ribosomal genes, which have been found to be especially good for differentiating species, were then compared. The DNA results from Lake A showed that the three lake microbes all clustered in the archaea part of the tree (the 0.05 scale bar in Figure 30 indicates a DNA sequence difference of 5 per cent), and that they were in close proximity to the species *Nitrosopumilis maritimus*, an archaeon that oxidizes ammonium to nitrite. The chemical habitat at 10–12m in Lake A (Figure 29) would be ideal for energy production based on ammonium oxidation, with oxygen diffusing down into the oxycline from above and ammonium diffusing up from the anoxic waters below.

DNA-based methods are proving to be valuable in exploring the most extreme of all polar aquatic habitats: subglacial lakes. These waters remain liquid throughout the year, but lie hundreds or thousands of metres beneath the Antarctic ice cap. The first of these was unexpectedly discovered under the research station Vostok, set up by the Soviet Union near the south geomagnetic pole during the International Geophysical Year (1957/8). Radio-echo sounding showed that the 3,750m-thick ice sheet at this location was underlain with liquid water to an astonishing 1,000m depth. Continued geophysical measurements revealed that this hidden water body, named Lake Vostok, is of vast extent, with an area of 14,000km^2 and an estimated volume of 5,400km^3. This quantity of water greatly exceeds many other large lakes of the world, such as Lake Ontario (1,640km^3). The discovery of an aquatic environment sequestered beneath the ice raised questions of great scientific as well as public interest: is Lake Vostok a tectonic basin of sterile water? Or could it be an active lake ecosystem that has somehow functioned for millions of years in the absence of life-giving sunlight?

The quest to look for life in Lake Vostok was especially motivated by astrobiologists, scientists who are interested in the origin, evolution, and limits of life on Earth, and in the conditions that might allow life to exist in places beyond Earth. Liquid water has been detected elsewhere in the Solar System, for example beneath the thick ice crusts of Europa, the smallest of the moons orbiting Jupiter, and Enceladus, the sixth largest moon of Saturn. Lake Vostok seemed like an appropriate analogue in considering the prospect of ecosystems in such places, and to develop the sterile, ice-penetrating technologies that would be needed to retrieve chemical and potentially biological samples from these environments. An even more compelling motivation came from the ongoing surveys of the Antarctic ice sheets, with the discovery that there are hundreds of subglacial lakes (mostly much smaller than Lake Vostok), and that many of these are connected in vast Amazonian-size basins of flowing waters, hidden beneath the thick ice. This would make subglacial waters one of the world's great ecosystem types, with a potentially large, downstream influence on the coastal Southern Ocean, into which these waters ultimately discharge. Geophysicists are also greatly interested in these subglacial liquid environments given that ice sheet stability and flow will also be affected by the presence of lubricating water at the ice–land interface, with effects on global climate, ocean circulation, and sea level.

The first attempts to sample these subglacial lakes were marred by setbacks and uncertainty. The Russians had already drilled to great depth at Vostok Station to obtain a record of past climate change. Their 3.4km-long ice core provided an unprecedented view of the natural cycles of greenhouse gas concentrations over the last 400,000 years, showing past maxima that we have now vastly exceeded through human activities. However, kerosene aircraft fuel had been used to keep the hole open over the ten years of drilling, and when the drill finally broke through the ice into the lake in February 2012, the resultant lake samples were possibly contaminated with this fluid, making it difficult to resolve any native microbial community. In December 2012, a British research team attempted to sample Lake Ellsworth, a 150m-deep subglacial waterbody in west Antarctica overlain by 3,400m of ice. They used a sterile, hot-water drilling system to ensure no contamination of the subglacial waters, but unfortunately the fuel supply ran out before drilling could progress beyond 300m depth into the ice.

In January 2013, an American team broke through into Lake Whillans in west Antarctica using hot-water drilling and a series of protocols to eliminate microbial and chemical contaminants. This lake is known to be an active system that fills and drains regularly, as indicated by shifts in its surface ice elevation, and at the time of sampling it was 2.2m deep and overlaid by 800m of ice. The team applied DNA sequencing methods to determine the microbial community structure and they found a diverse microbiome in the water, with a prevalence of ammonium-oxidizing archaea as in the oxycline of Lake A, and bacteria similar to nitrite oxidizers that have been isolated from Arctic permafrost. A sediment core was also taken from the lake and its microbial community included methane-consuming bacteria at the surface, and methane-producing archaea at depth.

Many questions remain, such as whether there is a microbial network of biological interactions with eukaryotic cells and viruses, and how representative is the Lake Whillans microbiome. However, these first results provided compelling evidence that the subglacial environment is a vast living ecosystem based on microbes that use inorganic chemicals for energy, along with other microbes that derive their energy from organic materials. Over the decades ahead, the exploration of Antarctic subglacial lakes will continue to be an exciting frontier for 'extreme limnology', and will also provide insights into how life survived beneath the vast ice sheets that covered much of the world during glacial epochs in the past.

Exploding lakes

At the opposite extreme to the cold water ecosystems of polar and alpine regions, geothermal waters also hold great interest for lake scientists and microbiologists. Here once again, life has been pushed to its limits of survival, yet a surprising variety of microbial extremophiles has the ability to thrive under these harsh conditions of low pH and scalding hot temperatures. The genomic techniques described above have been applied with great success to these waters, and have revealed unusual species and strategies for survival in these severe habitats. Some of the biomolecules found in the microbes living in these lakes have also proved to be of immense commercial value, and bioprospection of microbes from geothermal ecosystems has resulted in the development of novel products for use in biotechnology and in the biomedical industry. One of the most well known is the enzyme called 'Taq polymerase'. Taq is the abbreviation for *Thermus aquaticus*, a microbe first isolated from a hot pool at Yellowstone National Park, USA, and the original source of this enzyme that is used in an important technique for amplifying DNA for analysis, known as the polymerase chain reaction (PCR). *Thermus aquaticus* grows naturally at temperatures from 50 to 80°C, and its heat-stable DNA polymerase therefore proved ideal for the alternating high temperatures used in PCR.

A perennial hazard of living in an active geothermal region is that the ground and its associated waters have a tendency to blow up from time to time. This may be in the form of a hydrothermal explosion crater, where trapped gases including water vapour finally exceed the pressure resistance of their surrounding rock and soil, explode out of the ground, and leave behind a large hole that then fills with water to form a lake. The craters of active volcanoes may also fill with water that can be ejected from the lake during eruption, or drain through a breached wall of volcanic ash.

One such example is the crater lake on Mount Ruapehu, shown in Figure 31. The waters of this lake fluctuate greatly in temperature, with values up to 60°C, and a highly acidic pH that can be as low as 0.9. The volcano has experienced three major eruptions over the last 150 years, with smaller eruptions at more frequent intervals. The mountain is now carefully monitored and is installed with a seismic alarm system on its snowy slopes, which in case of eruption warns skiers to move rapidly to safety away from the slurry of lake water and sediments ('lahar') that could begin flowing down the valleys. This monitoring was prompted by the tragedy of 23 December 1953, when, following an eruption, Ruapehu crater lake burst its retaining dam of ash.

31. Highly acidic lake in the active volcanic crater of Mount Ruapehu, New Zealand.

The lahar flowed down a river gorge, collapsing a railway bridge on the main train line. Unaware of this sudden event that had happened only minutes earlier, the night express train and its first six carriages plunged into the gorge, killing 151 of the passengers onboard.

Volcanic lakes can pose other threats to human life, including as the result of their supersaturation of gases. Lake Nyos in Cameroon occupies the crater of an extinct volcano, but a magma chamber beneath the lake leaks carbon dioxide into the water, resulting in extreme concentrations that can be suddenly released at the surface due to landslides or earthquakes. A massive cloud of carbon dioxide was emitted from the lake in 1986, and suffocated 1,746 people and 3,500 livestock in the surrounding area. Since that time, tubes have been installed into the lake to vent the gases from its deep waters and to lower the risk of explosion. A similar accumulation of gas is known in Lake Monoun, also in Cameroon, where an eruption in 1984 released large quantities of carbon dioxide that suffocated and killed thirty-seven people.

A third, much larger, lake that is charged with volcanic gases is Lake Kivu, on the border of Rwanda and the Democratic Republic of Congo. The bottom waters of this large deep lake (2,700km^2; maximum depth 480m) interact with a volcano, which has resulted in the build-up of methane as well as carbon dioxide. These gases emanate from time to time from the lake, and pockets of toxic, CO_2-enriched air are locally called 'mazuku', a Swahili word meaning 'evil winds'. A silver lining to this dangerous situation is that the methane at depth is also a potential source of fuel for power generation. A plant has now been installed at the lake that pumps up water, and extracts and burns the methane. This generates around 26 megawatts (MW) of electricity, while also reducing the gas content and risk of catastrophic explosion of the deep waters of the lake.

Chapter 7
Lakes and us

> Humans exert a more powerful effect than any other animal on Nature and its inhabitants.
>
> F. A. Forel

When François Forel began his catalogue of the plants and animals of Lake Geneva, the first species he placed on the list was *Homo sapiens*. He introduced the notion that humans are not only part of the lake ecosystem through activities ranging from lake shore development to transport of goods and people (Figure 32), but also have the capacity to do great damage to a lake and its ability to provide essential services such as fisheries and safe drinking water. He observed that lake levels change through human intervention as well as natural causes, and he was an expert witness in litigation against the city of Geneva and its defective control structure at the outflow of the lake. Little did he realize the magnitude of dam-building and proliferation of artificial lakes that would take hold of human society in the 20th century and that continues with fervour today in developing countries.

Forel surveyed the vast expanse of Lake Geneva (area of 580km^2 and volume of 89km^3) with the sense that it would forever remain a limitless source of high-quality drinking water for all residents of the lake. Yet later in the 20th century, this lake, like so many

32. A traditional merchant vessel on Lake Geneva in the 19th century.

others around the world, began to experience the effects of eutrophication, with a rapid decline in water quality, depletion of its bottom water oxygen, and proliferation of algae. For Lake Geneva and all freshwater resources, the greatest challenges may lie ahead with global climate change and the associated impacts of increased temperatures, shifts in mixing patterns, extreme weather events, changing water supply, and altered habitat conditions for native and invasive species.

Dams large and small

For thousands of years, humankind has dammed and impounded waters to create artificial lakes and ponds. Up until the late 19th century, almost all of these were small in scale and included structures for crop irrigation, livestock watering, flood control, and domestic water supply, waterbodies for aesthetic and cultural purposes, millponds for water power, and impoundments for fish farming. Over the course of the 20th century, large-scale projects for navigation and hydroelectricity became symbols of progress, and brought considerable economic benefits along with a vast expansion of lake waters. The reservoirs of Europe currently total

100,000km^2 in area, including behind two large dams on the Volga River, the Kuybyshevskoye (6,450km^2) and Rybinskoye (4,450km^2) reservoirs. The World Register of Dams currently lists 58,519 'large dams', defined as those with a dam wall of 15m or higher; these collectively store 16,120km^3 of water, equivalent to 213 years of flow of Niagara Falls on the USA–Canada border. One of the largest hydroelectric schemes in the world is the James Bay complex in northern Quebec, Canada, which began operation during the late 1980s; this covers a total reservoir area of 11,800km^2 and has a current generating capacity of 16,500MW, with further expansions in progress.

Although dam-building has slowed or even reversed in the Western world, there is a new boom phase underway in Asia, Africa, and South America. The Three Gorges dam on the Yangtze River (1,084km^2; dam height 181m) in China began operation in 2012 and is now the largest hydroelectric power station in the world in terms of capacity (22,500MW). Around a hundred large dam projects are in advanced planning or construction in Africa, including the 145m-tall Grand Ethiopian Renaissance Dam on the Blue Nile River. More than 300 dams are planned or under construction in the Amazon Basin of South America, including the Belo Monte dam complex on the Xingu River.

Reservoirs have a number of distinguishing features relative to natural lakes. First, the shape ('morphometry') of their basins is rarely circular or oval, but instead is often dendritic, with a tree-like main stem and branches ramifying out into the submerged river valleys. Second, reservoirs typically have a high catchment area to lake area ratio, again reflecting their riverine origins. For natural lakes, this ratio is relatively low; for example this ratio for Windermere and Wastwater in the English Lake District is around 16; for Lake Geneva it is 13.8, while for Lake Tahoe the ratio is only 2.6, and it is a factor contributing to the long water residence time of this lake (650 years). In contrast, for Lake St-Charles, the dammed drinking water reservoir for Quebec City, the catchment to lake area ratio is 46; for Lake Mead on the Colorado River behind the Hoover Dam, USA, it is 640; and for the Three Gorges reservoir this ratio is 923. These proportionately large catchments mean that reservoirs have short water residence times, and water quality is much better than might be the case in the absence of this rapid flushing. Nonetheless, noxious algal blooms can develop and accumulate in isolated bays and side-arms, and downstream next to the dam itself.

Reservoirs typically experience water level fluctuations that are much larger and more rapid than in natural lakes, and this limits the development of littoral plants and animals. Another distinguishing feature of reservoirs is that they often show a longitudinal gradient of conditions. Upstream, the river section contains water that is flowing, turbulent, and well mixed; this then passes through a transition zone into the lake section up to the dam, which is often the deepest part of the lake and may be stratified and clearer due to decantation of land-derived particles. In some reservoirs, the water outflow is situated near the base of the dam within the hypolimnion, and this reduces the extent of oxygen depletion and nutrient build-up, while also providing cool water for fish and other animal communities below the dam. There is increasing attention being given to careful regulation of the timing and magnitude of dam outflows to maintain these downstream ecosystems.

Dams create new lakes for hydropower, irrigation, and drinking water, but the environmental costs are not always apparent at the time. Lake Urmia, a great salt lake (5,200km^2 at its maximum extent) in Iran that is renowned for its bird life has shrunk to 10 per cent of its original size, in part because its three primary inflows have been dammed for irrigation and hydropower. The resultant deposits of salt are blown around in the wind and affect farmlands as well as human health. Similar environmental problems are also encountered at the vast Aral Sea (Kazakhstan/Uzbekistan), which shrank from 68,000km^2 in the 1960s to around 7,000km^2

by 2005 as a result of inflow diversions for irrigation. A dam has now been built at the northern corner of this lake to retain water, dilute the salt, and restore the fishery in a small part of the original basin.

Dam-building can have wide ranging consequences for the original residents of the river basin, both human and animal. The Belo Monte project will flood and perturb lands used by thousands of Amazonian Indians, and the cultural impacts have generated international concern and protest. To build the Three Gorges Reservoir, some 1.2 million people were displaced, including the entire populations of thirteen cities. The dam is now an impediment to animal migration, including the Yangtze sturgeon and other endangered fish species, but the greatest effects may be downstream, with increased ship traffic and shifts in the annual flood regime. There is evidence that lower water levels in the Yangtze River floodplain due to reservoir operations are conducive to the transmission of parasitic flatworms from aquatic snails to humans, causing the severe disease 'snail fever' or schistosomiasis; a rising incidence of this disease has also accompanied other hydro-developments such as Egypt's Aswan dam. The lower water levels are putting wetlands at risk, and reducing the connections among habitats for fish and other aquatic animals. Perturbation of the flow regime may also affect the spawning, egg hatching, growth, and migration activities of native species that have evolved in response to the natural cycle of water fluctuation. The impacts of dams on fish biodiversity are of special concern in tropical river basins of the Amazon, Congo, and Mekong rivers, which currently hold around 4,200 species, of which 60 per cent are endemic. In total for these three basins, some 840 dams are currently operating or under construction, while another 445 are in various stages of planning.

The downstream effects of dams continue out into the sea, with the retention of sediments and nutrients in the reservoir leaving less available for export to marine food webs. This reduction can also lead to changes in shorelines, with a retreat of the coastal delta and intrusion of seawater because natural erosion processes can no longer be offset by resupply of sediments from upstream. Severe erosion has already been observed in the Yangtze River delta since the Three Gorges Dam began to operate. An additional effect of reservoirs is the mobilization of mercury from flooded vegetation and soils. This is methylated by bacteria to the more toxic form, methyl mercury, which can be further concentrated at each step of the food chain and transferred to downstream marine waters.

Many of us throughout the world depend on reservoirs for flood control, water supply, electricity, and economic well-being, and the ecosystem services provided by dammed lakes are now an integral part of our civilization. Dam-building continues in much of the world, and there are calls for an expansion of construction efforts to mitigate the effects of climate change on future water availability, reduce reliance upon fossil fuels, and keep up with the ever increasing needs of our global population, estimated to increase by another three billion over the course of this century. There are salutary reminders from the past, however, that the costs of such projects can often be underestimated, the benefits oversold, and the human and environmental consequences given inadequate thought and consideration, to the long-term detriment of social and ecological values.

Greening of the world's freshwaters

One of the most serious threats facing lakes throughout the world is the proliferation of algae and water plants caused by eutrophication, the overfertilization of waters with nutrients from human activities. This issue came to the fore in the mid-20th century with the realization that while lakes become gradually more nutrient-rich with time, lose their clarity, and eventually infill with sediment and plant growth, this slow, natural process can be hugely accelerated by increasing nutrient inputs from human activities in the surrounding catchments. The resultant nutrient-rich 'eutrophic' or (with even more enrichment) 'hypertrophic' waters are commonly referred to in the popular press as 'dead lakes'. The toxic algae and lack of oxygen in such lakes can result in death and extinction, however the term is a misnomer because eutrophic waters teem with aquatic life, but unfortunately dominated by noxious species that severely impair fishing, drinking water usage, and other ecosystem services.

Nutrient enrichment occurs both from 'point sources' of effluent discharged via pipes into the receiving waters, and 'nonpoint sources' such the runoff from roads and parking areas, agricultural lands, septic tank drainage fields, and terrain cleared of its nutrient- and water-absorbing vegetation. By the 1970s, even many of the world's larger lakes had begun to show worrying signs of deterioration from these sources of increasing enrichment. In Lake Geneva, for example, the winter Secchi depth plunged from the values measured by Forel of 15–20m in the 1870s, to at best 10m in the 1970s. Forel reported that the bottom waters of Lake Geneva were well oxygenated, even at 300m towards the end of stratification, but a hundred years later, deep water oxygen concentrations had fallen to hypoxic values of less than 2mg/L that excluded bottom-dwelling animals from certain parts of the lake and likely caused the extinction of some species such as the blind shrimp (Figure 22).

A sharp drop in water clarity is often among the first signs of eutrophication, although in forested areas this effect may be masked for many years by the greater absorption of light by the coloured organic materials that are dissolved within the lake water. A drop in oxygen levels in the bottom waters during stratification is another telltale indicator of eutrophication, with the eventual fall to oxygen-free (anoxic) conditions in these lower strata of the lake. However, the most striking impact with greatest effect on ecosystem services is the production of harmful algal blooms (HABs), specifically by cyanobacteria.

In eutrophic, temperate latitude waters, four genera of bloom-forming cyanobacteria are the usual offenders: *Microcystis*, *Dolichospermum* (formally known as *Anabaena*), *Aphanizomenon*, and *Planktothrix*. These may occur alone or in combination, and although each has its own idiosyncratic size, shape, and lifestyle, they have a number of impressive biological features in common. First and foremost, their cells are typically full of hydrophobic protein cases that exclude water and trap gases. These honeycombs of gas-filled chambers, called 'gas vesicles', reduce the density of the cells, allowing them to float up to the surface where there is light available for growth.

Put a drop of water from an algal bloom under a microscope and it will be immediately apparent that the individual cells are extremely small, and that the bloom itself is composed of billions of cells per litre of lake water. In the example shown in Figure 33, each cell is around 5μm in diameter, with a conspicuous bright spot caused by its gas vesicles that scatter the light. For such a tiny, solitary cell, its floating speed to the surface would be so slow as to be almost useless, but by combining that buoyancy in multicellular colonies, the flotation rate can be impressively fast, up to 5m per hour for *Microcystis* colonies.

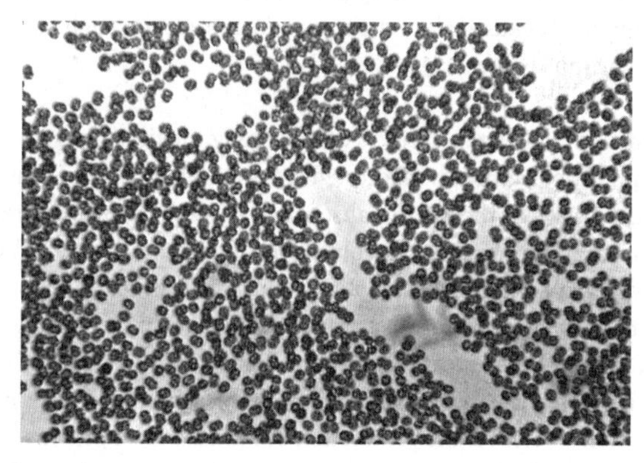

33. Photomicrograph of the toxic bloom-former *Microcystis aeruginosa*.

This flotation ability can also be regulated. During the day, the cells capture sunlight and produce sugars by photosynthesis; this increases their density, eventually to the point where they are heavier than the surrounding water and sink to more nutrient-rich conditions at depth in the water column or at the sediment surface. These sugars are depleted by cellular respiration, and this loss of ballast eventually results in cells becoming less dense than water and floating again towards the surface. This alternation of sinking and floating can result in large fluctuations in surface blooms over the twenty-four-hour cycle.

The accumulation of bloom-forming cyanobacteria at the surface gives rise to surface scums that then can be blown into bays and washed up onto beaches. These dense populations of colonies in the water column, and especially at the surface, can shade out bottom-dwelling water plants, as well as greatly reduce the amount of light for other phytoplankton species. The resultant 'cyanobacterial dominance' and loss of algal species diversity has negative implications for the aquatic food web, especially since these large colonial forms are difficult for the zooplankton to filter and ingest. Additionally, cyanobacteria tend to be deficient in essential fatty acids, and are therefore a poor quality food for animals. This negative impact on the food web may be compounded by the final collapse of the bloom and its decomposition, resulting in a major drawdown of oxygen.

Bloom-forming cyanobacteria are especially troublesome for the management of drinking water supplies. First, there is the overproduction of biomass, which results in a massive load of algal particles that can exceed the filtration capacity of a water treatment plant, especially if its intake is located at a depth or in a bay where these floating colonies accumulate. Second, there is an impact on the taste of the water. Cyanobacteria in general produce a broad spectrum of 'secondary compounds': biologically derived chemicals that do not participate in the primary processes of photosynthesis, respiration, and growth. For many, if not most of these compounds, it is not clear why cyanobacteria bother to produce them, although there is no shortage of hypotheses, ranging from chemical warfare on competing phytoplankton species and toxic dissuasion of herbivores, to the mobilization of trace metals and communication among cells. A number of these biochemicals produce unpleasant tastes and odours, including the musty, earthy odours of geosmin and 2-methyl isoborneol, the grassy taste of cyclocitrals, and sulfurous compounds such as alkyl sulfide that are released during decomposition. The third and most serious impact of cyanobacteria is that some of their secondary compounds are highly toxic.

Toxic lakes

On Saturday 2 August 2014, the mayor of Toledo, Ohio, held an emergency press conference. Mayor D. Michael Collins announced that residents should not drink or boil the tap water, and that all restaurants would be closed until further notice. The environmental chemists at the city's water treatment plant had detected a spike in a cyanobacterial toxin, microcystin-LR, during their routine testing of water quality, with values that exceeded the World Health Organization (WHO) limit of 1ppb. Toledo's water is drawn from Lake Erie, where cyanobacterial blooms occur each year over vast areas and often contain this toxin, sometimes at levels that exceed the WHO limit by a factor of 100. However, microcystin-LR is mostly retained within the cells and it can normally be removed by filtering off the algal particles. The problem at Toledo was that the toxin had made its way through to the post-treatment or 'polished' drinking water. After further tests and safety procedures, the tap water advisory was lifted the following Monday, but the closure of the water supply in a city of half a million people had lasting public impacts in the USA and Canada, and it refocused attention towards the seriousness of eutrophication and toxic water.

The issue of cyanobacterial toxins (cyanotoxins) has also been of great concern in other parts of the world. Lake Taihu ('Great Lake' in Chinese) is the third largest lake in China and is the drinking water supply to ten million people. Although vast in size (2,338km^2), it is shallow (maximum depth 2.6m) and highly eutrophic, and experiences a continuous bloom of *Microcystis aeruginosa* throughout the year. In 2007, residents of the city of Wuxi shifted to bottled water for a month because of the strange tastes and odours in their tap water, and the possibility of imbibing cyanotoxins from the lake. These concerns continue to this day, with efforts to improve monitoring and pollution control for this important water resource.

Microcystins are a class of water soluble toxins produced by many species of bloom-forming cyanobacteria, but most notably by the cosmopolitan species *Microcystis aerginosa*. Chemically they are classified as peptides: each molecule is a series of amino acids linked together with peptide bonds, just like proteins. Unlike many proteins, however, they are not denatured by boiling, possibly because the amino acids are arranged in a stable ring structure (Figure 34). This sturdy configuration also resists the effects of protein degrading enzymes (proteases) that are exuded by bacterial decomposers. Detailed analytical research has shown

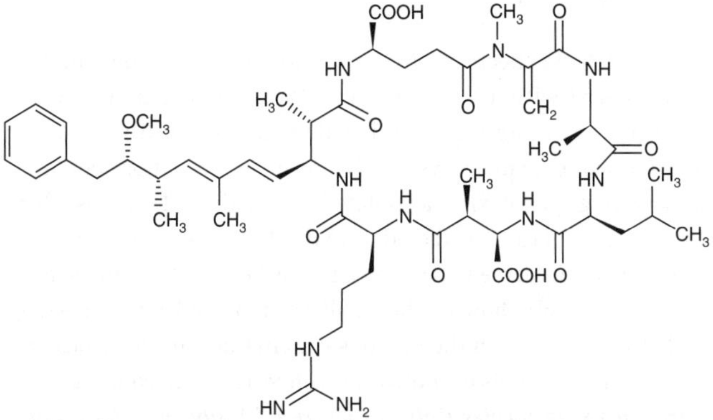

34. The toxic peptide microcystin-LR produced by cyanobacteria.

that although the basic ring structure is the same, the side groups can vary greatly, and more than a hundred chemical variants or 'congeners' of microcystins have been detected in *Microcystis*-rich waters, with microcystin-LR the most toxic.

The resistance of microcystins to water treatment protocols belies their biochemical reactivity. Once inside mammals, these toxins move to the liver where they are taken up by cells and interrupt the activity of key enzymes, specifically phosphatases. This ultimately results in liver damage, with associated oxidative stress to the kidney, brain, and reproductive organs. There is also evidence that microcystins have carcinogenic effects by disrupting microtubule assembly and cellular division. Nausea, vomiting, and gastrointestinal illnesses have been linked to drinking water containing microcystins, but the only known human fatalities were reported from a hospital in Caruaru, Brazil, in 1996. More than a hundred renal patients became ill, and seventy of them died during dialysis sessions with water derived from a reservoir with a *Microcystis* bloom; microcystins were detected in the water purification system of the clinic and also in samples of blood and liver of the patients. There are also many reports of sickness and deaths of dogs and farm animals that have drunk water containing toxic cyanobacterial blooms, including *Microcystis*.

When the toxic effects of *Microcystis* first became known in the 1950s, assays with mice showed that the cyanotoxins caused rapid mortality and they were labelled the 'Fast Death Factor', now known to be microcystins. However, an even more potent, faster acting cyanotoxin was isolated in Canada in the 1960s after several herds of cattle had died from drinking bloom-containing waters, and this came to be known as the 'Very Fast Death Factor'. It was ultimately shown to be an alkaloid, named 'anatoxin-a', with potent effects on the nervous system that can cause death within minutes. This cyanotoxin was first isolated from the nitrogen-fixing species *Dolichospermum* (*Anabaena*) *flos-aquae* that is commonly found in eutrophic lakes, but it is now known to be produced by several other species and genera of cyanobacteria.

In addition to microcystins and anatoxin-a, bloom-forming cyanobacteria produce a variety of other compounds that are toxic to wildlife, domestic animals, and humans. These include organophosphates, bioactive amino acids, and paralytic shellfish toxins. Some species produce cell wall materials and other compounds that cause skin irritation and dermatitis, and there are many cases of skin allergy reactions by swimmers in bloom-infested waters. Some of these reports, however, may be due to another cause: larval flatworms (schistosomes) that infect freshwater snails and ducks, but that can also burrow into human skin to cause 'swimmer's itch'.

Clearing the water

Can a eutrophic 'dead lake' be restored to its original, near-pristine condition? This important but ambitious goal requires an understanding of the mechanisms and processes that lead to the overproduction of water plants and algae, especially toxic cyanobacteria. In the latter half of the 20th century, when many lakes of the world were experiencing the effects of rapid population growth and increasing discharge of pollutants into their waters, the discussions revolved around three nutrients: carbon, nitrogen, and phosphorus. The North American soap and detergent industry was reluctant to see any mandated changes to their production of phosphate-rich products, and argued that carbon was the element causing eutrophication. Short-term experiments with bottles of lake water enriched with carbon seemed to support this argument, although some of these bioassays gave equivocal and contradictory results.

The most convincing evidence of how nutrients cause eutrophication and which element often plays the greatest role in lakes came from the work of Canadian limnologist David W. Schindler in the

Experimental Lakes Area (ELA) of Canada. The vast granite lands of northern Canada called the Precambrian Shield contain millions of lakes and ponds that have been scratched out from the rock by recent glacial activity, and a small area of this lake-rich landscape in northern Ontario was set aside in 1968 for whole-lake observations and experiments. Schindler's experiment was elegant in its simplicity, and it produced results at the whole ecosystem level rather than in an artificial laboratory environment. He and his team installed a nylon-reinforced vinyl curtain across the middle of an hourglass-shaped lake (number 226 in the ELA inventory) and one side, the southwest basin, was fertilized with carbon (as sucrose that would be rapidly converted to carbon dioxide by the bacteria) and nitrogen (as nitrate). The other side, the northeast basin, was also fertilized with carbon and nitrogen, but additionally with the third element phosphorus (as phosphate), all in ratios approximating those in discharges from sewage treatment plants.

The results were spectacular (Figure 35). The carbon plus nitrogen enriched side of the lake showed little change in algal biomass, as measured by the photosynthetic pigment chlorophyll. This was especially interesting in that Lake 226, like other Canadian Shield lakes, had low natural levels of dissolved inorganic carbon; if carbon were to have an enrichment effect, this would be one of the best places to look for it. In contrast, the lake on the carbon plus nitrogen plus phosphorus side of the curtain soon developed a noxious algal bloom dominated by nitrogen-fixing cyanobacteria. This bloom gave the water a turbid green appearance, and the Secchi depth shrank from around 3m to 1m. Apart from the differences in many of the water quality variables that were measured, the striking visual contrast in the lake between the two sides of the curtain provided compelling evidence to policy makers: phosphorus is the key nutrient limiting bloom development, and efforts to preserve and rehabilitate freshwaters should pay specific attention to controlling the input of phosphorus via point and nonpoint discharges to lakes.

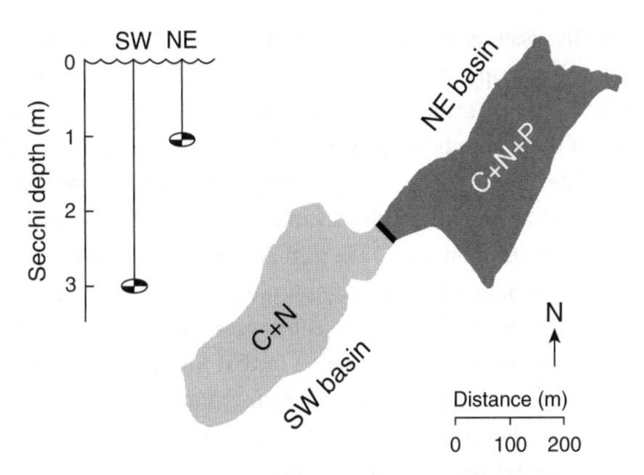

35. Cyanobacterial bloom development in the CNP-fertilized northeast basin of Lake 226 and the resultant drop in Secchi depth.

Over the last half century, there has been a focus of attention throughout the world on controlling phosphorus by identifying sources, moving effluent out of the lake basin, installing phosphorus-stripping systems, and regulating the use of phosphorus-rich detergents and other products. One of the earliest examples was Lake Washington, USA, where sewage effluents from the burgeoning city of Seattle were discharged into the lake at ever increasing rates, up to 80 million litres per day in the early 1960s. Work by W. Tommy Edmondson at the University of Washington drew attention to the precipitous decline in water quality of the lake including an increase in nutrients and proliferation of cyanobacteria, and these findings ultimately led to a diversion scheme to redirect the city effluents to the sea. This was implemented in a step-wise fashion over a period of several years, and by 1968 there was no sewage effluent discharged into the lake. Over the five-year period from 1964 to 1969, Edmondson's team showed that there was massive improvement in water quality, with mean algal concentrations in summer declining by a factor of 6, in tandem with a similar decline in winter phosphate concentrations.

Current discussions now focus on the question: is it enough to control only phosphorus? Carbon is in plentiful supply in lakes from the overlying atmosphere and from inorganic and organic sources in the watershed. However, there are multiple reasons for also considering nitrogen. One counter-argument to targeting nitrogen is that nitrogen-fixing cyanobacteria, such as the species that rose to prominence in Lake 226 after carbon + nitrogen + phosphorus enrichment, have an unlimited source of nitrogen that cannot be controlled: gaseous nitrogen in the atmosphere. However, this is not entirely correct in that nitrogen fixers derive only part of their nitrogen requirements from the atmosphere and must otherwise depend on forms such as ammonia, nitrate, and organic nitrogen in the water. Furthermore, one of the most worrisome of bloom-forming cyanobacteria in lakes and reservoirs is *Microcystis*, and this species is unable to fix nitrogen. It produces the nitrogen-rich toxin microcystin (Figure 34; each molecule has ten atoms of N), and there is evidence that the production of this toxin is enhanced by nitrogen enrichment. This toxic species appears to be undergoing a resurgence throughout the world, stimulated by nitrogen- as well as phosphorus-rich fertilizers that spill from agricultural lands out into waterways though increasingly efficient soil drainage systems.

Certain lakes throughout the world lie on catchments that are naturally rich in phosphorus, and are strikingly different in their chemistry from those of the ELA (e.g. lakes on the central volcanic plateau of the North Island, New Zealand, and Lake Titicaca in South America). For these lakes, full control of phosphorus loading is unrealistic. Unbalanced phosphorus control may also lead to macrophyte expansion, since these plants have access through their roots to the bountiful reserves of phosphorus in the sediments, while benefiting from the ongoing nitrogen-enrichment of the overlying water. Finally, freshwater lakes ultimately discharge into the sea, and coastal marine environments tend to be rich in phosphorus, with limiting concentrations of nitrogen. To consider only phosphorus removal may transfer the problem of eutrophication downstream to these receiving waters.

For all of these reasons, the Environmental Protection Agency of the USA and the European Union have recommended control of nitrogen as well as phosphorus removal from effluents. This policy decision has been controversial because nitrogen-removal treatments are expensive and technically more difficult than phosphorus-stripping. The focus upon a single element phosphorus provided a clear, unambiguous target that policy makers and managers could focus upon, and in the process the loading of all nutrients has often been reduced, for example by piping treated effluent out of the basin (e.g. Lake Tahoe and Lake Washington), or by the use of natural and engineered wetlands to remove nitrogen as well as phosphorus. Whatever the local decision, Schindler's results from Lake 226 and his related experiments at the ELA will always be compelling evidence that humans have the capacity to rapidly shift lakes from pristine waters to noxious blooms, and that control of external nutrient supply is essential to preserve our lakes from algal overproduction.

Lake water recovery after nutrient controls have been put in place has not always been as rapid or as complete as was hoped for. Part of the problem is that of 'hysteresis': the trajectory of a lake during its recovery phase after nutrient reductions may follow a different path to the one that led to its degradation, particularly if the lake has undergone a 'regime shift' to persistent noxious blooms, sharply decreased water transparency, and, as a result, loss of underwater plant communities such as algal charophytes (Figure 36). There are many processes that affect this return pathway and that cause a slow-down or inertia to restoration measures. Part of this may be for biological reasons. For example, after many years of cyanobacterial growth, the sediments may contain abundant resting spores and dormant cells that are then an inoculum for ongoing blooms. However, one of the greatest effects causing the slow pace of recovery is nutrient release from

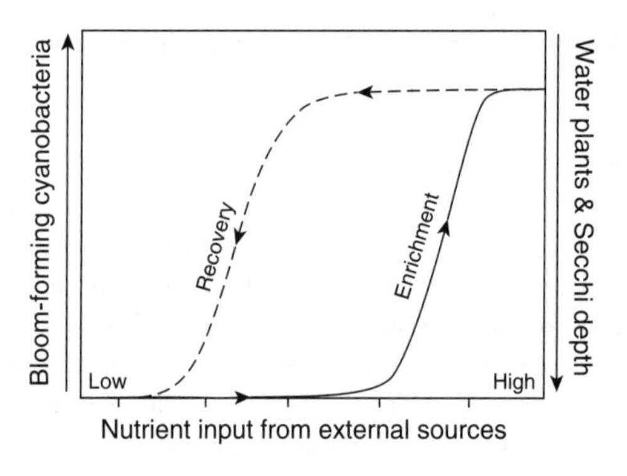

36. Hysteresis in the recovery versus degradation of a lake.

the sediments, especially phosphorus, which can accelerate under anoxic conditions (with known exceptions). This is referred to as 'internal loading' to separate it from the external loading on the lake from its surrounding catchment. Such effects are a concern for Lake Erie, for example, where the severe toxic blooms are the result of increases in external loading but are compounded by internal phosphorus release from anoxic sediments (Figure 19). More release means more algal growth, producing more biomass for decomposition and more oxygen depletion towards anoxia. This is a vicious circle that is difficult to stop, and by far the best approach is to protect our lake waters before they reach this oxygen-depleted state.

The future of lakes

In his monograph on Lake Geneva, Forel drew attention to the importance of physical, chemical, biological, and human features of the lake, and the need to incorporate all of these aspects into an integrated synthesis, 'an overview of all the detailed facts, where each specialised study would be supported by the data from other studies'. This type of integrated overview is today at the heart of Earth system science, in which each aspect of the environment, from geophysics to human processes, is considered an interacting part of the global system. An integrated, system-level perspective on lakes is now vitally relevant to managing the world's freshwater resources in the face of rapid climate change at a planetary scale.

Lakes at many sites throughout the world now show evidence of warming, on average at rates that approximate the increases in air temperature. However, there are large variations in the extent of this warming, even among lakes in the same climate region because of differences in depth, wind exposure, and transparency. Extreme weather conditions are likely to accompany climate warming, and in parts of Europe and North America heavy rainfall events have been identified as a factor causing increased inputs of coloured dissolved organic matter and a resultant 'browning' of lakes. This organic enrichment can modify aquatic food webs (Figure 16), and the reduction of transparency means that more solar energy is absorbed in the surface waters, further contributing to warmer temperatures.

Increases in water temperature can have wide ranging effects that propagate throughout the lake ecosystem. Evaporation rates increase with warming, and this may shift the water balance towards net loss, and a reduction in lake levels that is either offset or exacerbated by changes in rainfall. Even small fluctuations in water level can have serious effects on important ecological features. For example, the wetlands around the North American Great Lakes are important for migratory birds and many fish species, and these semi-aquatic habitats are especially vulnerable to minor shifts in water balance. Rising temperatures will also reduce the extent of favourable habitats for cold-water specialists among the flora and fauna, and will facilitate invasion by species from warmer climates.

One of the less obvious effects of climate change in lakes is on the layering (stratification) of the water, with warmer conditions at the surface producing a greater difference in temperature and therefore density between the surface and bottom waters. This stronger stratification provides more resistance to mixing by the wind, and it lessens the exchange of oxygen from the atmosphere to depth in the lake as well as the transfer of nutrients from deep waters to the surface. In Lake Tanganyika, this effect of warming and increased stability appears to have resulted in decreased nutrient supply to the photic zone by mixing, resulting in decreased phytoplankton production and a 30 per cent drop in fish yield. These thermal stratification effects are of special concern for managing noxious blooms of cyanobacteria, which are directly stimulated by warm temperatures and also prefer stable waters for their migration based on gas vesicles and buoyancy control.

Forel described how the majority of residents of the Lake Geneva basin lived in close proximity to the lakeshore and were a dependent part of the lake ecosystem. This latter idea ran counter to the prevailing view at the time and through much of the 20th century that humankind lives on a higher plane than Nature, with full dominion over the land, the air, and the water, and unlimited license to exploit these resources to meet our ever growing needs. At the time that Forel was writing the third volume of his monograph and describing the human history of Lake Geneva, some 56,000 people lived at Lausanne, the world population was around 1.6 billion and atmospheric CO_2 levels were 296ppm; over the subsequent hundred years, these local and global populations both increased by a factor of four, and CO_2 levels rose by 25 per cent. Today, more than 800,000 people draw their water from the lake, and there is ongoing concern about controlling nutrient inputs and contaminants, including emerging pollutants such as pharmaceuticals, microplastics (polyethylene particles less than 5mm in size), and engineered metal nanoparticles (1–100nm) that are increasingly common in the world's freshwaters. Additionally, and like many lakes elsewhere, Lake Geneva has begun to show the effects of climate change, with evidence of warming bottom waters, changes in stratification and mixing, and shifts in the spawning dates of some fish species.

The increasing impacts of population growth and global change on lakes throughout the world are a reminder that while we may be the most powerful entities in the biosphere, we have a close reciprocal relationship with our planetary environment, and a vested interest in protecting its integrity and the ecological services that we crucially depend upon. Lakes are centres of biodiversity, reactive 'slow rivers' that flow, mix, and counter-flow, conduits to the atmosphere and ocean, integrators of the surrounding environment, and sentinels of change in the past and present. From flood control and transport systems to reservoirs of water, food, and energy, they are key resources for human society. To protect and sustain all of these values will require policy decisions and action at a global level, as well as ongoing advances in lake science and local management practices, and attention to the integrative approach that is the hallmark of 'limnology'.

Further reading

Historical and literary

G. Bachelard, *Water and Dreams: An Essay on the Imagination of Matter* (Dallas: The Pegasus Foundation, 1983).

C. Bertola, *Léman Maniac* [*Crazy about Lake Geneva*] (Nyon: Éditions Glénat, 2009).

J. Dennis, *The Living Great Lakes: Searching for the Heart of the Inland Seas* (New York: Thomas Dunne Books, 2003).

D. Egan, *The Death and Life of the Great Lakes* (New York: W. W. Norton & Company, 2017).

F. N. Egerton, 'History of Ecological Sciences, Part 50: Formalizing Limnology, 1870s to 1920s', *The Bulletin of the Ecological Society of America* 95(2): 131–53 (2014).

F. A. Forel, 'Notice sur l'histoire naturelle du lac Léman' [Notes on the Natural History of Lake Geneva], pp. 217–43 in: E. Rambert, H. Lebert, Ch. Dufour, F. A. Forel, and S. Chavannes (eds), *Montreux* (Neuchâtel: H. Furrer, 1877).

F. A. Forel, 'Allgemeine Biologie eines Suesswassersees' ['General Biology of a Freshwater Lake'], pp. 1–26 in: O. Zacharias (ed.), *Die Tier- und Pflanzenwelt des Suesswassers* [*The Flora and Fauna of Freshwaters*] (Leipzig: J. J. Weber, 1891).

F. A. Forel, *Le Léman: Monographie limnologique* [*Lake Geneva: Limnological Monograph*], Vols I, II, III (Lausanne: F. Rouge & Compagnie, 1892, 1895, 1904).

F. D. C. Forel (ed.), *Forel et le Léman: Aux sources de la limnologie* [*Forel and Lake Geneva: To the Origins of Limnology*] (Lausanne: Presses Polytechniques et Universitaires Romandes, 2012).

J. B. Gidmark, *Encyclopedia of American Literature of the Sea and Great Lakes* (Westport: Greenwood Press, 2001).

W. Grady (ed.), *Dark Waters Dancing to a Breeze: A Literary Companion to Rivers and Lakes* (Vancouver: Greystone Books, 2007).

B. Green, *Water, Ice & Stone: Science and Memory on the Antarctic Lakes* (New York: Harmony Books, 1995). A captivating, insightful account of lake science in the field.

J. Hart, *Storm over Mono: The Mono Lake Battle and the California Water Future* (Berkeley: University of California Press, 1996). This book has inspired students to become environmental scientists.

J. Kirk, *In the Domain of the Lake Monsters* (Toronto: Key Porter Books, 1998).

R. L. Lindeman, 'Seasonal Food-Cycle Dynamics in a Senescent Lake', *American Midland Naturalist* 1: 636–73 (1941).

S. Plath, *Crossing the Water* (London: Faber & Faber, 1975).

A. W. Reed, *Treasury of Maori Folklore* (Wellington: A. H. & A. W. Reed, 1963).

A. Steleanu, *Geschichte der Limnologie und ihrer Grundlagen* [*History of Limnology and its Foundations*] (Frankfurt: Haag & Herchen, 1989).

S. Tesson, *The Consolations of the Forest: Alone in a Cabin on the Siberian Tundra* (New York: Rizzoli International Publications Inc., 2013). A modern-day *Walden* set at Lake Baikal, Russia.

A. Thienemann, 'Seetypen' ['Lake Types'] *Naturwissenschaften* 9: 343–6 (1921).

H. D. Thoreau, *Walden* (New Haven: Yale University Press, 2006). This fully annotated, affordable version of Thoreau's 1854 classic is edited by Jeffrey S. Cramer, curator of the Thoreau Institute.

G. Topping (ed.), *Great Salt Lake: An Anthology* (Logan: Utah State University Press, 2003).

M. Twain, *Roughing It* (New York: Harper and Brothers, 1872). Includes entertaining accounts of Mark Twain's visits to Lake Tahoe and Mono Lake.

W. F. Vincent and C. Bertola, 'Lake Physics to Ecosystem Services: Forel and the Origins of Limnology', *Limnology and Oceanography e-Lectures*, 4(3), doi:10:4319/lol.2014.wvincent.cbertola.8 (2014). Available at: <http://www.cen.ulaval.ca/warwickvincent/PDFfiles/303-Forel.pdf>.

Popular guides to lake science and aquatic biology

M. J. Burgis and P. Morris, *The World of Lakes: Lakes of the World* (Ambleside: Freshwater Biological Association, 2007).

D. Gilpin and J. Schmid-Araya, *The Illustrated World Encyclopedia of Freshwater Fish & River Creatures* (London: Hermes House, 2009).

U. Lemmin, *Voyage dans les abysses du Léman* [*Voyage into the Abyssal Depths of Lake Geneva*] (Lausanne: Presses Polytechniques et Universitaires Romandes, 2016).

B. Moss, *Ponds and Small Lak*es: *Microorganisms and Freshwater Ecology* (Exeter: Pelagic Publishing, 2017).

L.-H. Olsen, J. Sunesen, and B. V. Pedersen, *Small Freshwater Creatures* (Oxford: Oxford University Press, 2001).

G. K. Reid et al., *Pond Life: A Guide to Common Plants and Animals of North American Ponds and Lakes* (New York: St Martin's Press, 2001).

D. W. Schindler and J. R. Vallentyne, *The Algal Bowl: Overfertilization of the World's Freshwaters and Estuaries* (Edmonton: The University of Alberta Press, 2008).

Scientific books

J. L. Awange and O. Ong'ang'a, *Lake Victoria: Ecology, Resources, Environment* (Heidelberg: Springer, 2006).

T. D. Brock, *A Eutrophic Lake: Lake Mendota, Wisconsin* (New York: Springer Verlag, 1985).

C. Brönmark and L.-A. Hansson, *The Biology of Lakes and Ponds* (Oxford: Oxford University Press, 2005).

G. A. Cole and P. E. Weihe, *Textbook of Limnology* (Long Grove: Waveland Press, 2016).

W. K. Dodds and M. R. Whiles, *Freshwater Ecology: Concepts and Environmental Applications of Limnology*, 2nd edn (San Diego: Academic Press, 2010).

S. I. Dodson, *Introduction to Limnology* (New York: McGraw-Hill, 2005).

J.-C. Druart and G. Balvay, *Le Léman et sa vie microscopique* [*Lake Geneva and its Microscopic Life*] (Versailles: Éditions Quae, 2007).

A. J. Horne and C. R. Goldman, *Limnology* (New York: McGraw-Hill, 1994).

J. Kalff, *Limnology: Inland Water Ecosystems* (Upper Saddle River: Prentice Hall, 2002).

G. E. Likens (ed.), *Encyclopedia of Inland Waters*, 3 volumes (Oxford: Elsevier, 2009).

B. Moss, *Ecology of Freshwaters: A View for the Twenty-First Century*, 4th edn (Oxford: Wiley-Blackwell, 2010).

S. T. Ross, *Ecology of North American Freshwater Fishes* (Berkeley: University of California Press, 2013).

J. P. Smol, *Pollution of Lakes and Rivers: A Paleoenvironmental Perspective*, 2nd edn (New York: John Wiley & Sons, 2008).

R. W. Sterner and J. J. Elser, *Ecological Stoichiometry: The Biology of Elements from Molecules to the Biosphere* (Princeton: Princeton University Press, 2002).

J. H. Thorp and A. P. Covich (eds), *Ecology and Classification of North American Freshwater Invertebrates*, 3rd edn (Oxford: Elsevier, 2010).

J. G. Tundisi and T. M. Tundisi, *Limnology* (Boca Raton: CRC Press, 2012).

W. F. Vincent and J. Laybourn-Parry (eds), *Polar Lakes and Rivers: Limnology of Arctic and Antarctic Aquatic Ecosystems* (Oxford: Oxford University Press, 2008).

J. D. Wehr, R. G. Sheath, and J. P. Kociolek (eds), *Freshwater Algae of North America: Ecology and Classification* (San Diego: Elsevier, 2015).

R. G. Wetzel, *Limnology: Lake and River Ecosystems*, 3rd edn (New York: Academic Press, 2001).

Scientific articles and reviews

S. Bonilla and F. R. Pick, 'Freshwater Bloom-Forming Cyanobacteria and Anthropogenic Change', *Limnology and Oceanography e-Lectures* 7(2) (2017), <https://doi.org/10.1002/loe2.10006>.

J. Catalan et al., 'High Mountain Lakes: Extreme Habitats and Witnesses of Environmental Changes', *Limnetica* 25: 551–84 (2006).

B. C. Christner, J. C. Priscu, et al., 'A Microbial Ecosystem beneath the West Antarctic Ice Sheet', *Nature* 512: 310–13 (2014).

J. J. Cole et al. 'Plumbing the Global Carbon Cycle: Integrating Inland Waters into the Terrestrial Carbon Budget', *Ecosystems* 10: 172–85 (2007).

B. R. Deemer et al., 'Greenhouse Gas Emissions from Reservoir Water Surfaces: A New Global Synthesis', *BioScience* 66: 949–64 (2016).

J. A. Downing, 'Emerging Global Role of Small Lakes and Ponds', *Limnetica* 29: 9–24 (2010).

D. Dudgeon et al. 'Freshwater Biodiversity: Importance, Threats, Status and Conservation Challenges', *Biological Reviews* 81(2): 163–82 (2006).

L. Grattan and V. Trainer (eds), 'Harmful Algal Blooms and Public Health', *Harmful Algae* 57B: 1–56 (2016).

F. Hölker et al., 'Tube-Dwelling Invertebrates: Tiny Ecosystem Engineers have Large Effects in Lake Ecosystems', *Ecological Monographs* 85(3): 333–51 (2015).

S. MacIntyre and R. Jellison, 'Nutrient Fluxes from Upwelling and Enhanced Turbulence at the Top of the Pycnocline in Mono Lake, California', *Hydrobiologia* 466: 13–29 (2001).

M. V. Moore et al., 'Climate Change and the World's "Sacred Sea"—Lake Baikal, Siberia', *BioScience* 59: 405–17 (2009).

R. J. Newton et al., 'A Guide to the Natural History of Freshwater Lake Bacteria', *Microbiology and Molecular Biology Reviews* 75: 14–49 (2011).

C. M. O'Reilly et al., 'Rapid and Highly Variable Warming of Lake Surface Waters around the Globe', *Geophysical Research Letters* 42(24) (2015).

H. W. Paerl et al., 'It Takes Two to Tango: When and Where Dual Nutrient (N & P) Reductions are Needed to Protect Lakes and Downstream Ecosystems', *Environmental Science & Technology* 50: 10805–13 (2016).

L. G. M. Pettersson, R. H. Henchman, and A. Nilsson, 'Water: The Most Anomalous Liquid', *Chemical Reviews* 116: 7459–62 (2016).

S. Pointing et al., 'Quantifying Human Impact on Earth's Microbiome', *Nature Microbiology* 1: 16145 (2016).

D. Righton et al., 'Empirical Observations of the Spawning Migration of European Eels: The Long and Dangerous Road to the Sargasso Sea', *Science Advances* 2: e1501694 (2016).

J. P. Smol, 'Paleolimnology: An Introduction to Approaches Used to Track Long-Term Environmental Changes Using Lake Sediments', *Limnology and Oceanography e-Lectures* 1(3) (2009), <https://doi.org/10.4319/lol.2009.jsmol.3>.

J. A. Stenson, 'Differential Predation by Fish on Two Species of *Chaoborus* (Diptera, Chaoboridae)', *Oikos* 31: 98–101 (1978).

J. D. Stockwell et al., 'Habitat Coupling in a Large Lake System: Delivery of an Energy Subsidy by an Offshore Planktivore to the Nearshore Zone of Lake Superior', *Freshwater Biology* 59: 1197–212 (2014).
C. A. Suttle, 'Environmental Microbiology: Viral Diversity on the Global Stage', *Nature Microbiology* 1: 16205 (2016).
C. S. Turney and H. Brown, 'Catastrophic Early Holocene Sea Level Rise, Human Migration and the Neolithic Transition in Europe', *Quaternary Science Reviews* 26: 2036–41 (2007).
C. Verpoorter et al., 'A Global Inventory of Lakes Based on High-Resolution Satellite Imagery', *Geophysical Research Letters* 41: 6396–402 (2014).
C. E. Williamson et al., 'Ecological Consequences of Long-Term Browning in Lakes', *Scientific Reports* 5: 18666 (2015).
K. O. Winemiller et al., 'Balancing Hydropower and Biodiversity in the Amazon, Congo, and Mekong', *Science* 351: 128–9 (2016).

Websites

ASLO: <http://aslo.org>.
Billow demonstration (Kelvin–Helmholtz instabilities): <https://www.youtube.com/watch?v=UbAfvcaYrOO>
Chironomid tubes (video): <https://www.youtube.com/watch?v=RQwau_uSyy4>
Colour of lakes: <http://www.citclops.eu/transparency/measuring-water-transparency>
Cyanobacteria-identification: <https://pubs.usgs.gov/of/2015/1164/ofr20151164.pdf>
Dams (data base): <http://www.icold-cigb.org/GB/world_register/world_register_of_dams.asp>
Daphnia feeding (video): <https://www.youtube.com/watch?v=pLL_YzZ_4O0>
Daphnia swimming (video): <https://www.youtube.com/watch?v=MyDQ_f1mzH8>
English Lake District and other U.K. lakes: <https://eip.ceh.ac.uk/apps/lakes/>
Kelvin waves: <https://www.youtube.com/watch?v=SZlix47Jq4A>
Lake Biwa: <http://www.biwahaku.jp/english/member-e/researchactivities.html>
Lake Tahoe: <http://terc.ucdavis.edu/>
Large lakes (IAGLR site): <http://www.iaglr.org/lakes/>

Léman/Lake Geneva—International Commission (CIPEL) <http://www.cipel.org/>

Methane from Arctic thaw lakes (video): <https://www.dailymotion.com/video/x2mwrcv>

Microbial mats in Lake Untersee, Antarctica (video): <https://www.youtube.com/watch?v=qs2hUZP-6Bo>

Mono Lake: <http://www.monolake.org/>

NALMS: <https://www.nalms.org/>

Phantom midge (video): <https://www.youtube.com/watch?v=LQCj6T5sdQM>

Plankton and benthos (images from the Freshwater Biological Association): <http://www.environmentdata.org/browse-collection>

Rotifers (images): <http://www.microscopy-uk.org.uk/mag/wimsmall/extra/rotif.html>

SIL: <http://limnology.org/>

Tubifex worms—sludge worms (video): <https://www.youtube.com/watch?v=hxYBiBi3EbE>

World Lake Database (reference catalogue): <http://wldb.ilec.or.jp/>

Index

A

B

C

D

E

F

G

H

I

J

K

L

M

N

S

T

U

V

W

Z